Ch 4939

CC
635 782

Ath Arena

THE COVERED GARDEN

THE COVERED GARDEN

BY
KENNETH LEMMON

Foreword by
THE DUKE OF DEVONSHIRE

LONDON
MUSEUM PRESS LIMMED

First published in Great Britain by Museum Press Limited
26 Old Brompton Road, London, S.W.7
1962

FOR STUART

PRINTED IN GREAT BRITAIN BY
BUTLER AND TANNER LTD., FROME AND LONDON
R.3327

FOREWORD

I am very proud to write a brief foreword on Mr. Lemmon's book *The Covered Garden*. Knowing as I do both the author's ability as a writer and his knowledge and love of gardening, I am sure that all who read his book will be richly rewarded.

The subject is one close to my own heart as Chatsworth in former days had one of the finest covered gardens in Sir Joseph Paxton's Great Conservatory.

All lovers of greenhouse gardening will I know be deeply indebted to Mr. Lemmon for this scholarly and fascinating book.

Chatsworth, July, 1961   DEVONSHIRE

PREFACE

THE "jungles under glass" of the old stove conservatories of the Victorian era have always had a peculiar fascination for me, tied up, maybe, with a boyhood's avid reading of tales of travel and adventure on far-off foreign shores. So, when I came to gardening and to greenhouse work as a keen amateur and no more, I wished to read all about conservatories, greenhouses, winter gardens, or covered gardens as they were often called. I did not want to know how to manage a greenhouse, what to grow in one or how to grow it. I wanted instead to read of their evolution: what had prompted men to attempt to force the blooms of spring from a winter clime, how had technical progress assisted the gardener, where had all the strange plants come from, who brought them, why had the old glory gone and what had prompted that truly magnificent era of Victorian "glassiloquence"?

I found but scattered reference here and there, and there seemed to be a gap in our gardening histories. There and then I was prompted to fill it and started a most intriguing course of research which has led to the attempt for the first time to put between the covers of one book the colourful history of "The Covered Garden."

I can only hope my readers will find the story as fascinating reading as I have found its writing and that their gardening under glass will be the more rewarding for their knowledge of the long traditions, skills and glories of the past and the exciting possibilities of the future.

My sincere thanks are due in no small measure to the great band of "curious gardeners" of yesteryear and a host of practical ones of today whose writings and whose talk respectively have inspired and encouraged me to face the task of getting together, however modestly, this story of the covered garden and greenhouse history and practice through the ages.

For practical help in many ways my warmest thanks must go

to Mr. F. Beckwith, the most kindly and patient librarian of the Leeds Library, to Mr. A. B. Craven for introducing me to the Gott Collection of Leeds Reference Library, to Mr. A. Simmons and the staffs of the Royal Horticultural Society's Lindley Library, to Miss Bertha Backhouse, a colleague, for most valuable assistance with the typescript and lastly to Dr. D. G. Ineson, Ll.B., F.L.S., for many suggestions, for reading and checking the manuscript and for his warm and kindly encouragement at all times.

Morley, June, 1961   KENNETH LEMMON

ACKNOWLEDGEMENTS

THE Author and Publishers wish to express their thanks for permission to reproduce photographic illustrations as follows: No. 2, Royal Botanical Gardens, Kew; Nos. 13 & 14, Honeywell Controls Limited, Greenford, Middlesex.

CONTENTS

ILLUSTRATIONS

DRAWINGS IN THE TEXT

9

ILLUSTRATIONS

HALF-TONE ILLUSTRATIONS
(Between pages 136 and 137)

I

FIRST BEGINNINGS

As man's material wealth and leisure increases, as his civilisation reaches the high peak where his way of life and his material surroundings become malleable and flexible in his hands, it is then that he adds the grace-notes of living; the artist and the craftsman flourish and all that is best in that civilisation comes to the top.

While artists work with enjoyment and skill at their various crafts to reflect their life and age, to gardens and gardeners is added the flourish of floriculture as distinct from agriculture, horticulture or arboriculture. Man returns to the soil for pleasure and recreation; instead of toiling he potters; no longer a tiller—he is a gardener.

The last refinement of gardening which comes with a wealthy, leisured oligarchy of taste and elegance is the desire to have and to cultivate all the exotic treasures of the floral world under artificial conditions irrespective of climate or locality.

So it was that the nineteenth century saw in Britain a greenhouse and covered garden tradition housing a vast wealth of tropical greenhouse exotics surpassing in the highest degree any such tradition or floral richness ever seen in the world before or since.

It was a slow climb to the heights of that greenhouse century, but always it was the material obstacles that held up progress—unsuitability and non-availability of constructional materials, wrong techniques, difficulties of communication and travel, the vastness of unexplored territories—never the human element. For from the 1500's in England (and from before the Christian era in Southern Europe) there is evidence of man's questing for the unusual in the floral calendar and, among the few who had the means, there was ever present the desire to grow in their own domains the riches of a tropic world, so little known and

11

understood but so full of magic and of romance to make a man's heart ache to possess such faerie gold; to cosset in a pot a rich natural jewel of the East.

The greenhouse story is, then, a story of slow but inevitable progress which for this country follows the social pattern so closely as to be an integral part of it; so closely indeed that one can speak with reasonable accuracy of the Orangery age as that of the noble dilettantes of the Restoration, the Stove era as that of the philosophical, experimental period—the Age of Elegance; and that of the Greenhouse and Conservatory as that of the highly decorative, social status period—the Victorian.

But at the outset let us look at the "glass" roots of this tradition, let us see where this strange idea of surrounding oneself with the opulence of floral nature had its birth.

In delving into the past in an attempt to establish the accurate date, the place, or the personalities or personality connected with some radical change in the technique or philosophy of one or other of the natural sciences, the research worker nearly always comes up against some almost insoluble problem. When a change is epoch-making and tangible, as the invention of the steam-engine or the splitting of the atom, there is irrefutable evidence, definite dates, and clearly identifiable and known individuals, all detailed in the histories of the times. But it is different when the change, although ultimately it may be revolutionary and striking, is gradual and piecemeal as when an attempt is made to find the original pioneers of the greenhouse.

For while it could only be in our Northern climate of cold winters, when there flourished a wealthy aristocracy with the necessary leisure and the inclination, that any great greenhouse tradition could arise, there is ample evidence to show that the Romans were well aware of the advantages of "glass" and heat, and used it for the protection and forcing of fruit, including the grape, and vegetables out of season.

The Greeks in a minor way were also conversant with the practice. It is thought by inference and mentioned by M'Intosh that "The Gardens of Adonis" were forcing-houses in miniature.

He quotes Plato in his *Phaedon* as saying that "a grain of seed or the branch of a tree placed in or introduced to these gardens, acquired in eight days a development which cannot be obtained in as many months in the open air."

The gardens, it is reasonably to be assumed from various authorities, were rather like the miniature gardens seen at flower shows in these days, often made up by children. Branches of flowering trees, grain or bulbous subjects were put in pots or earthenware dishes and a form of bell-jar placed over them and then put on the flat roofs in the sun to bring the contents to maturity in time for the great fertility feast of Adonis in which the gardens formed such an important part of the ritual.

Columella, the Roman writer on rural matters, implying some form of protection, states: "Rome possesses within the precincts of her walls fragrant trees, trees of precious perfumes such as grow in the open air of India and Arabia. . . ." But it devolved upon Tiberius Caesar to bring a utilitarian note to these early attempts at greenhouse work.

Told by his doctor that his illness needed a cucumber a day for its cure, his gardener was called upon to "dispense" the prescription. He showed that the cucumber could be grown *fere toto anno*. This was done in pits filled with fermenting dung covered with frames of mica or talc.

Columella added: "Anyone who wishes to have the cucumber ripe earlier should, when mid-winter is passed, procure well-manured soil enclosed in baskets and given a small amount of water. Then, when the seeds have come up, he should place the baskets in the open air on warm and sunny days in a building so that they may be protected from any blasts of wind. But, if it is cold and stormy, he should bring them back under cover and continue to do so until the spring equinox is over. He should then sink the whole baskets into the ground. He will then have early fruits. It is also possible, if it be worth the trouble, for wheels to be put under the larger vessels so that they may be brought out with less labour. In any case, the vessels ought to be covered with slabs of transparent stone, so that in cold weather when the days are clear, they may be brought into the sun. By this method Tiberius Caesar was supplied with cucumbers during almost the whole year."

Sir Joseph Banks, who remarked on dessert fruit forcing in Roman times in the Horticultural Society *Transactions*, claimed that the forcing was done by means of *specularia* (stoves) and that the "glass" was "Muscovy glass" (*lapis specularis*), actually sheets of mica split thinly, and that this must have been so as this method of glazing windows was widely used in Rome. Indeed, the Roman plumber and glazier was called *specularius*.

In these early greenhouses it seems that peaches were attempted, for the epigrammatists of the day, including Martial, spent not a little of their wit and satire to ridicule the idea, making play with the word "pale," suggesting that these early greenhouse peaches were but the palest green and not much better than turnips to eat.

M'Intosh puts his faith in Seneca's evidence on the Roman forcing-house and quotes the writer as saying: "Do not those live contrary to nature who require a rose in winter and who, by the excitement of hot water and an appropriate modification of heat, force from winter the later blooms of spring?"

In the ruins of Pompeii was found a building which archaeologists have called a *specularia*. It has tiers of masonry for displaying plants, hot-air flues in the walls, and every indication of once having possessed a frontage of talc or rough glass.

Curiously enough the record seems to stop there, although probably it is not strange after all. The trouble of running a forcing-house, however rudimentary, in the warmer sunny climate of Southern Europe could not really have been worth while after the initial novelty had worn thin.

Be that as it may, the fact remains that there was little or no advance to report from Southern Europe in *specularia* or greenhouse construction, beyond those already noticed, for hundreds of years. Combing through the tomes of gardening history in general, which, as opposed to greenhouse history in particular, has many, many volumes to its credit, one finds the first recognisable attempt at early greenhouse practice in Western Europe came from Albertus Magnus, a Paduan who in 1259 is said to have entertained William of Holland, King of the Romans, in a garden maintained in flower and fruit by artificial heat. The only other reference in these early centuries is that in a letter signed "John," dated 1385, stating that in "the Bois de Duc in France they grow flowers in glass pavilions turned to the South."

Pontanus (1426–1503) in 1490 mentions that coal was being used to over-winter citrus plants in wooden sheds, and at Padua in Italy about 1550 Daniel Barbaro ordered a *viridarium* or greenhouse to be built in the botanical gardens there, to which plants could be taken during the winter. This would almost certainly have been a wooden, brick or stone storeroom with a fire hearth or a portable brazier. Two writers of the sixteenth century, C. Estienne and Jean Liebault, mention that citrus plants were wintered in arched cellars in Europe, while in 1599 a plant shelter was built for the Leyden Botanical Gardens. It was called the *ambulacrum*, indicating by its very name the type of structure it must have been; obviously a shed-type orangery in which plants were placed in rows to form a parade during the winter months.

Nearer at hand in time and certainly in documentation is Salomon de Caus who in 1619 at Heidelberg used the first moveable wooden structure, said to have been glazed with glass, to shelter 400 orange-trees belonging to the Elector Palatine. A drawing of the shelter shows it to have been a Dutch barn type of building opened up for the summer, and closed in for the winter by wooden shutters and roofing. In the walls small window openings were few and far between. De Caus, who was both architect and engineer to the Elector, wrote about the orangery himself in a publication dated 1620 wherein he said: " . . . there are 30 large orange-trees, each about 25 foot high and above 400 others both middling sized and small. These orange-trees are about 60 years old. The orangery is 280 foot long and 32 foot wide and it is made of wood which is put up every year about Michaelmas and the orange trees are warmed by means of four furnaces all the winter, so that in the time of the great frosts one can walk in this orangery without feeling any cold on account of the heat which proceeds from the furnaces. At Easter the framework is taken away to leave the trees uncovered all the summer. On account of the expense of keeping it in repair I advised His Majesty to have the orangery covered with freestone so that in winter the roof being already covered in, it was only necessary to close the windows, and thus much woodwork might be spared."

The wealthy merchants of Venice and Genoa had also been busy introducing new and unusual plants from foreign climes to the more genial air of Italy. Some of the floral wealth and colour of the East had been captured and brought over with the silks and spices to grace the villa gardens, and these garden treasures, along with the fair scented orange, had not passed unnoticed by the Englishman travelling abroad, who now came across flowers and plants he had never seen before, nor even realised existed, whose beauty of form, colour or maybe their curious and unusual structure, gave rise to his wish to grow such novel and beautiful plants in his own garden at home.

Not until the late sixteenth century in England, however, had a taste grown for exotic flowering plants so that it could be said of Gerard's collection both at his master Lord Burleigh's garden and at his own physic garden in Holborn that: "Now it may appear that our ground would produce other fruit besides hips and haws, acorns and pignuts."

At that time even kitchen vegetables were imported from France and Holland!

Pulteney pays tribute to these early collectors of exotics whose enthusiasm led to a specialised greenhouse cult in the following century because of the need for plant shelter. Mentioning Gerard, he praises also his assistant James Garet, a London apothecary, who was also a lover of exotics and an enthusiastic tulip-raiser. Nicholas Lete, also of London, was noted as a man "in love with rare and fair flowers for which he does send into Syria, having a servant in Aleppo and in many other countries."

Lord Edward Zouch at Hackney had de l'Obel, the Flemish botanist, as his gardener and brought plants and seeds from Constantinople. Lord Hansdon, Lord High Chamberlain, was worthy of triple honour, his contemporaries said, for his care in getting, as also his care in keeping, such rare and strange things from the farthest parts of the world. But the citrus fruits and particularly the orange were the principal reason that brought the greenhouse ultimately to Britain, and not the herbs and simples of Gerard or the tulips of Garet, nor even the "rare and strange things" of Lord Hansdon, which were almost certainly more herbs and simples which the hot dung-bed and frame of either oiled paper or thick green glass could well take care of if they came from warmer climes.

Early attempts at protection and conserving during the winter are found described in good, down-to-earth gardening language by Gervase Markham and the great John Parkinson of the *Paradisi*.

Parkinson, writing on orchards in 1629 says: "I bring to your consideration the orenge alone without mentioning citron or lemmon trees, in regard of the experience we have seen made of them in divers places. For the orenge tree hath abiden with some extraordinary looking [after it] and tending of it, when as neither of the other would by any means be preserved any long time. They must be kept in great square boxes and lift there to and fro by iron hooks in the sides . . . to place them in an house or close gallery in for the winter time . . . but no tent or mean provision will preserve them."

Markham, a Nottinghamshire man, said to have been the first to import an Arab horse to this country, in his work *The Whole Art of Husbandry* of 1631 was very insistent on the new secret discovered. "I have seen," he said, "divers noblemen and gentlemen which have been very curious in these dainty flowers [he was referring among others to the daffodil, the hyacinth, narcissus and tulip] who have made large frames of wood with boards of 20 inches deep standing upon little round wheels of wood, which being made square or round, according to the master's fancy, they have been filled with choice earth the most proper to the flower they would grow and place them in the garden where they may have the strength and violence of the sunshine heat all day

B

and the comfort of such moderate showers and at night draw them by men's strength into some low vaulted gallery joyning upon the garden where they may stand warm and safe from storms, winds, frosts, divers blastings and other mischief which ever happens in the sun's absence, and in this manner you may not only have of many of dainty outlandish flowers, but also all sorts of the most delicatest fruits that may be as the orange, limond, pomegranate, cynamon tree, olive and almonde."

Lest anyone might think that was but an imaginary supposition, said Gervase Markham, "I can assure him that within seven miles of London the experiment is to be seen where all these fruits and flowers with a world of others are growing in two gardens most abundantly."

Probably the first reference to hot-houses, as opposed to orangeries or winter plant-shelters, in this country's literature is in Barnaby Googe's *Four Books of Husbandrie*, published in 1578, but Johnson, who cites the instance, considers that these hot-houses can have been only but rough wooden sheds warmed by a crude stove or hearth, and that, on the evidence available, would seem to be a correct interpretation, for Worlidge in *Systema Horti-culturae* or the art of gardening, published in 1677 does not mention the subject at all.

Yet it is obvious when we reach the seventeenth century that things were happening and had happened, outside our own British Isles, to stimulate men of wealth and taste to an interest in things horticultural and in things exotic. Men were travelling further afield. America was part of the background. Europe was nearer home, and more like home, for the noblemen and courtiers of a Charles who had spent so much time in France and the Lowlands.

An intriguing reference to an early hot-house occurs in Nicholas Breton's *Fantasticks* of 1626, where he tells his readers to beware of hot-houses for fear of catching cold; but Hazlitt, who remarks the passage, said no one could quite make up their minds whether the hot-houses mentioned were of the horticultural kind, the London gaming houses or those houses of ill repute in London called the Stews!

Francis Bacon noted the early attempts at protection in his celebrated essay "On Gardens," and he, too, as early as 1597, mentions hot-houses, *en passant*, so that we can gain no idea what type of house he was referring to. But it is fair to infer some form of artificial protection under sheds, accompanied by heat, as the following extract from his essay shows:

" . . . For December and the latter part of November you must take such things as are green all winter—holly, ivy, bays, juniper, cypress trees, yew, pineapple trees, fir trees, rosemary, lavender, periwinkle, the white, the purple and the blue germander, flags, orange trees, lemon trees and myrtles, if they be stoved, and sweet marjoram warm set."

In Holland, Leyden Botanical Gardens had advanced from its early *ambulacrum*, and, between 1680 and 1687, the first planthouses were built. Plant introduction was impossible without warmed glasshouses and from the *dedicatio* of Hermann's catalogue of 1687 it is apparent that the garden could boast of this necessary equipment, although the evidence points to buildings with opaque roofs and glass frame fronts with more glass than the masonry-fronted *ambulacrum*.

In 1644 John Evelyn saw the orangeries at Cardinal Richelieu's château at Rueil and noted a new name for these structures when he wrote: " . . . this leads to the *citronière* where there is a very noble conserve of all those rarities."

Evelyn, particularly, occupies an honoured niche in this history of the covered garden for he was the first English writer to use the words "greenhouse" and "conservatory." It is true that he used both names synonymously, but he is credited with the coining of both words by the *Oxford English Dictionary*, and I am sure the honour is rightly bestowed.

John Rea in his *Flora, Ceres et Pomona* of 1676 has an interesting point or two to make about the attempts in his time to preserve the few exotics which gardeners had to deal with. Rea himself was a great lover of horticultural rarities. "By long continued diligence," he said, "over 40 years of alegiance to that lovely recreation I have collected all those rare plants, fruits and

An early Dutch orangery with decorated back wall and small lattice
window; used for a winter promenade. Note the free-standing Dutch
stove requiring fuelling inside the house.

(*Den Nederlandtsen Hovenier*, J. van der Groen, Amsterdam, 1670.)

flowers that by any means I could procure either in this nation France or Flanders."

John Rea used hot-beds to raise and preserve many of his rarities, but he also experimented with other methods to over-winter his rare treasures, which must have pleased his patron Sir Thomas Hanmer, Bt., who loved novelty and skill so much that Rea dedicated his book to him, "To the truly noble and perfect lover of ingenuity."

Grapes ripened too late in this country, he said, unless defended with a tilt or mat; an early vinery in embryo! Like so many people of his day and age he grew citrus-trees and like his fellow-gardeners he had to have somewhere to keep them from harm during winter.

So, John Rea, with the practicality so characteristic of his age, noted that at the side of the nursery beds there "should be a con-venient house to put in such necessary tools as are to be useful about the garden, but chiefly for housing your greens and other tender plants in winter, for which purpose it ought to have a stove or raised hearths in several places, that with a small fire you may gently attemper the air in time of hard frosts. Rose bays and oranges are more tender and must be planted in strong cases to be housed in winter. . . . But shut them not up in day time especially, unless constrained by fogs or frosts which last long. You must on fairer days acquaint them again with the sun and air by degrees."

But Rea was not quite so sure about the efficiency of his toolhouse cum conservatory for he has a further word about times of extreme frost "when water will freeze in your conservatory." Then he advised in default of stoves or raised hearths "you must attemper the air with pans of charcole especially at night. Let the cole be half burned out before the pans be placed and then set them not too near the plants."

For those without a greenhouse or a conservatory he ad-vocated digging a hole in the ground, under a wall, five feet deep, bricked round, and with a wall a foot high above the ground on which boards would rest. The plants were then to be plunged underground, care being taken that when the sun shone the boards were raised like a trapdoor. Even with these crude methods and the use of hot-beds and frames, some success was attained for there is an authentic account of an installation dinner given by Charles II at Windsor on 23 April, 1667, when the tables bore—in April, mind you—cherries and strawberries with ice-cream.

Sir William Temple's *Upon the Gardens of Epicurus, or of gardening in the year 1685,* notes that orangeries were becoming a popular feature of the gardens of his many rich and influential friends. He wrote of the times—"so mightily improved [is gardening] in the three or four and twenty years of His Majesty's reign that perhaps few countries are before us either in the elegance of our gardens or in the number of our plants. . . . My orange trees are as large as any I saw when I was young in France, except those at Fontainebleau, or what I have seen in the Low Countries, except some very old ones of the Prince of Orange, as laden with flowers as any can be, as full of fruit as I suffer or desire them to be, as well tasted as any that are brought over except the best sorts of Seville or Portugal."

John Evelyn, who it will be remembered had first written about orangeries in 1644, had become almost a consultant on the new cult of conservatory gardening that was being tried by his many wealthy friends, relatives and neighbours. He told of the Carews' garden at Beddington where in 1580 some of the first orange-trees seen in England were planted. They were in the open ground in the summer and "secured in winter only by a tabernacle of boards and stoves. . . ."

In 1668 he was busy advising Lord Sandwich of the faults of the then stove-houses and told that patron of Pepys that if only the stoves (stove-houses) "could be lined with cork, I believe it would better secure them from the cold and moisture of the walls than either mattresses or reeds with which we commonly cover them." He was right, of course; but the reeds—what a haven for pests, mildew and all we would not want in the stove-house!

Advising the same gentleman on the proper treatment of pomegranates, he has but a poor opinion to give of their treatment in the heated structure of those days, for he says he had always kept the fruit exposed "and the severest of our winters does it no prejudice. They will flower plentifully, but bear no fruit with us, either kept in cases in the repository, or set in the open air; at least very trifling, with the greatest industry of stoves and other artifices." I think we can take the other artifices to be the hot-bed and the oiled cloth very often wrapped round a wooden frame in some attempt to keep these importations from the worst buffetings of the cold and inclemency of the English climate.

At a later date Evelyn writes to Lord Sandwich that he questions the science of the construction of the heating apparatus of the day, and to tell him of the science of regulating heat by use of the proper fuel, saying that if certain classes of coal, heavily charged with sulphur, were burned then no delicate vegetables could exist in the atmosphere. At Chelsea, he said, they had the remedy, "they keep the doors and windows of the houses open."

At Woburn the Earl of Bedford was an early grower of oranges, and records of the household round about the year 1671 show a picture, noticed later by Celia Fiennes:

"Passing through the orchards and kitchen gardens towards the house," the description ran, "an archway at the end of the cherry orchards led on to a terrace running immediately under the west front of the house on to which the windows of the state rooms looked. Here, in the summer, were kept potted trees of oranges, lemons, myrtles and aloes, which, in the winter, were either removed into the house, or into the new orangery especially arranged for their reception where they might enjoy the warmth of a fire, or are placed in troughs and covered with boards for their better protection."

John Field, who later became one of the famous partnership with George London, at the Brompton Nurseries, was the Earl's gardener and in his accounts and records of the time there is mention of stoves for the orangeries, brought over from Holland. George London procured them for Field, and it is recorded that at the time there was some little dispute about the price of stoves which arose over the proper rate of exchange between the Dutch guilder and the English currency.

Sir Hugh Plat, a man of invention and resource, a characteristic product of his age, wrote of forcing and conservatory practice in 1660 in *The Garden of Eden*. He tells of Sir Francis Carew of Beddington who, when Queen Elizabeth called on him, had late cherries to give her. He had covered one of his trees with canvas kept damp to retard the fruit so that it would be just ripe when the Queen arrived and managed to retard it for a month.

He spoke also of Mr. Jacob, a glass manufacturer, saying: "I have known Mr. Jacob of the Glassehouse to have carnations all winter by the benefit of a room that was near his glassehouse fire." He also mentions gillyflowers as a fit subject for forcing and asks: "*Quaere*, If pease, beans, pompeons, musk mellons and other pulse seeds put in small pots . . . and placed in a gentle stove or some convenient place aptly warmed by a fire and then sown in March or April would they come up sooner?"

He also asks: "Why not utilize a kitchen fire planting them [apricots or vines] near a warm wall, or brewers, diers, soap boilers or refiners of sugar, who have continual fires, can easily convey the heat of steam of their fires (which are now utterly lost) into some private room adjoining, wherein to bestow their fruit trees?"

It was very obvious that Sir Hugh was keenly interested in the subject and had spent much time and thought in pondering it. For instance another of his suggestions was to grow plants and fruit against concave walls lined with lead or tin to cause reflection that "they might happily bear their fruit in our cold clymate." He goes on to add that if these walls were convenient to kitchen fires to warm them at the back that also might add materially to the ripening of the fruit.

In France there was progress, too, encouraged by the Court's need for early fruits and salads. At Versailles in 1678 Jean de Quintinye started to build a hot-house and finished it in 1683, in which to force fruit and vegetables, mainly strawberries, cherries and salad crops.

In 1685, also at Versailles, Mansart completed for Louis XIV the noblest orangery in Europe. This architectural wonder, with its three arcaded galleries all opaque roofed, was of heroic dimensions with its central façade 508 feet long, 42 feet wide by 45 feet high, and each of the other flanking galleries 375 feet long. It could hold its winter store of 1,200 orange-trees as well as 300 other kinds of tender shrubs.

Fagon, Superintendent of the Jardin Royal in Paris for Louis XIV about 1693, constructed hot-houses, it is claimed, with glass roofs, which he "warmed with stoves and furnaces" for the preservation of tender plants, among them the tea shrub.

As good a picture of the exotic gardening of the day in England as can be seen anywhere, occurs in the pages of Celia Fiennes' journal as she rode side-saddle around the country from about 1694 until 1703. Covering hundreds of miles under atrocious conditions, with roads but bogs, and paths but pitfalls by the way, this intrepid lady was as appreciative of a good garden, its design and its innovations in the way of plants, hot-houses, orangeries or the newest grottoes, as the keenest professional gardener of the day, and her descriptions are as detailed as they are pleasant to read in their bridging of the gardening centuries. One of her first rides was from London into Oxfordshire and to Oxford where she visited the Physic Garden there in 1694. One can do no better than allow her to speak:

"The physic garden at Oxford afforded great diversion and pleasure, the variety of flowers and plants would have entertained one a week, the few remarkable things I took notice of was the aloes plant which is like a great flag in shape, leaves and coullour, and grows in the form of an open hartichoke, and towards the bottom of each leafe it is very broad and thicke in which there are hollows or receptacles for the aloes; there is also the Sensible plant, take but a leafe between finger and thumb and squeeze it and it immediately curles up together as if pained, and after some tyme opens broad again, it looks in coullour like a filbert leafe but much narrower one inch long; there is the Humble plant [*Mimosa pudica*] that grows on a slender stalke and do but strike it, it falls flatt on the ground stalke and all, and after some tyme revives againe and stands up, but these are nice plants and are kept mostly under glass, the aire being too rough for them."

While still in Oxford she visited New College and noticed that the serviteurs of the college could live very neatly, as she phrased it, "if sober, and have all their curiosityes; they take much delight in greens of all sorts, myrtle, orange and lemons and lorrestine [*laurestinus*] growing in pots of earth, and so moved about from place to place and into the aire sometymes."

At Breamore the seat of Lady Brooks she saw for the first time an orange-tree and at Hinchinbrooke, Lord Sandwich's home, she noted that the gardens and wilderness and greenhouse would be very fine when they were quite finished.

A different way of dealing with the oranges and exotics she saw at Bretby in Derbyshire, the home of the Earls of Chesterfield, where, writing of the gardens she says: " . . . Beyond this garden is a row of orange and lemon trees set in the ground, of a man's height and pretty big, full of flowers and some large fruit almost ripe; this has a penthouse over it which is covered up very close in the winter: this leads on to a great wilderness and just by it is another square with a fountaine whose brim is decked with flower potts full of flowers and all sorts of greens, on either side is two or three rows of orange and lemon trees in boxes one below the other in growth."

At the Crown Inn in Whitchurch in Shropshire she noticed a fine array of "orange and lemon trees, mirtle striped and gilded holly trees," while at Epsom, at Sir Thomas Cooke's house, she observed hot-beds with convenient houses near by and walks set with dwarf trees, cypress and myrtle.

Towards the end of the 1600's a Queen's scholar and Fellow at Trinity College, Cambridge, and master of Enfield (Middlesex) Grammar School, Dr. Robert Uvedale, was said to be one of the earliest owners of a hot-house in the country. A contemporary, writing of his gardens towards the end of the century, in 1691 to be precise, was able to say of both them and the doctor: "Dr. Uvedale is a great lover of plants and having an extraordinary art of managing them, is become master of the greatest and choicest collection of exotic greens that is perhaps anywhere in the land. His greens take up six or seven houses or roomsteads. His orange trees and largest myrtles fill up his biggest house and another house is filled with myrtles of a less size, and those more nice and curious plants that need closer keeping are in warmer rooms, and some of them stoved when he thinks fit. His flowers are choice, his stock numerous and his culture of them very methodical and curious."

At Chatsworth in 1697 the first Duke of Devonshire built a greenhouse with an arcaded front and, from Kip's drawing, apparently with an opaque roof. But it looked a handsome building and was unique at the time in such a setting to be labelled "greenhouse" rather than orangery. This house was pulled down by the third Duke in 1749, but the old original front was preserved and rebuilt into the plant-house one can still see at Chatsworth. And at nearby Wollaton Hall in Nottinghamshire an orangery said to have a glass roof and the first of its kind, was built in 1696; according to Loudon it was designed by those royal gardener professionals, London and Wise, but accurate historical records are non-existent.

2

ORANGERIES AND HOT-HOUSES

CERTAINLY in Europe and in this country it was the need for conservation of the orange-tree which proved the stimulant for the development of shelter-houses which were both more efficient and ornamental than the old tabernacles of boards with their charcoal hearths sunk in their dank earth or paved floors. The orange had to be preserved during English winters; it was a tremendously popular garden subject, hence the rise of the orangery, the conservatory and the greenhouse—and some two hundred years elapsed in British garden history before gardeners made much distinction between the three.

Many were the gardeners who spoke highly in praise of the orange. Olivier de Serres round about 1663 wrote: "It is impossible to express the great beauty of these precious plants, proceeding from the unmatched and dazzling colour of their foliage and the excellent qualities of their fruits which, contrary to the nature of all others, remain attached to the trees for the greater part of the year; and what augments their claim is that we see at one and the same time on the same stem, little and middle-sized and large, and see their flowers for a while accompanying them, breathing a delicious fragrance in the place where they are shut up."

It was possible at the height of their popularity to have the choice of 169 sorts.

There was considerable evidence of the difficulties of growing citrus before both gardeners and garden architects got to grips with the problem and experience, as always, had taught many a disappointing lesson. Rembert Dodoens, the old Dutch herbalist, in his herbal of 1569 spoke of the orange which would only give fruit if kept away from the cold in winter in the house. Jean Liebault, writing in 1573, advocated the placing of citrus plants in arched cellars during the winter. The royal gardeners in France had their own methods, no good at all for perambulating among during the winter, but successful only in conservation of the plants.

The art of the French in this way, it was noted, was their ability to shut up in barn-like buildings or cellars, orange-trees, camellias and other evergreens without light, warmth, air or watering for three months in winter. The secret, it was said, was the French gardener's practice of witholding water towards the time of hibernation, ripening the wood and seeing that the plants were in a dormant state before shutting up. They were gradually inured to the open and the elements by the process in reverse.

The task of conserving citrus plants in Great Britain was indeed fraught with disappointment and difficulty, but it was a challenge and led indirectly to the betterment of all shelter-houses.

In this country the honours must go to Lord Burghley as a pioneer in the growing of citrus plants and as being one of the first Englishmen to feel the need of a structure for winter plant protection as distinct from tilt-mats or hot-beds. At Burghley Court he built round about 1561 what was probably the first orangery, for he was writing to Paris in March of that year asking for lemons and pomegranates to go with an orange-tree he already had. A letter to Mr. Thomas Windebank who was in Paris at the time told that gentleman that he (Lord Burghley) had heard from his son Thomas that Mr. Carew was going to have certain trees sent home and added; "I already have an orange tree, and if the prise be not too much I pray you procure for me a lemon, a pomegranate and a myrtle tree; and help that they may be sent home to London with Mr. Carew's trees; and beforehand send me in writing a perfect declaration how they ought to be used, kept and ordered."

The answer, dated 8 April, 1562, told his Lordship that the writer had managed to get the myrtle-trees in two pots at a cost of a crown each and a lemon-tree—15 crowns. They were to be planted in cases (plant-tubs) and stood out in some sheltered place in the garden for the summer, only to be lifted into the house from September until April. At Burghley Court a large room called the Orange Court is still attached to the house, where it is said these plants were preserved during the winters.

A detailed impression of one of the early orangeries is given in a Parliamentary survey of November, 1649, of the orange garden at Wimbledon belonging to Henrietta Maria, Queen to Charles I. It runs: " . . . in the side of which said Oringe Garden there stands one large Garden House, the out walls of brick; fitted for the keeping of Oringe trees; neatly covered with blue slate, and ridged and guttered with lead; the materials of which house, with the great door and iron thereof, with a certain stone pavement lying before those doors, in nature of a little walk four foot wide and seventy-nine foot long, we value to be worth £66 13s. 4d.

"In the said Garden House there are now standing in square boxes, fitted for that purpose, fourty-two Oringe trees bearing fair and large oringes; which trees, with the boxes and the earth and materials therein feeding the same, we value at ten pounds a tree, one tree with another, in toto amounting to £420.

"In the said Garden House there now also is one Lemon tree bearing great and very large lemons, which, together with the box that it is grown in, and the earth and materials therein feeding the same, we value at three pounds a tree, one with another, in toto £18."

Eighteen "oringe" trees which had not then borne fruit were valued at £90.

Yet the plant remained a novelty and for the pleasure of the few, for in 1666 we find Pepys making a delightfully Pepysian diary note. It was on 25 June that the good Mr. Pepys went to Hackney. He wrote; "Mrs. Pen carried us to two gardens at Hackney (which I every day grew more in love with) Mr. Drake's one, where the garden is good, and house and prospects admirable, the other my Lord Brooke's where the gardens are much better but the house not so good nor prospect good at all. But the gardens are excellent, and here I first saw oranges grow, some green, some half, some quarter, some full ripe, on the same tree, and one fruit of the same tree do come a year or two after the other; I pulled off a little one by stealth (the man being mightily curious of them) and eat it, and it was just as other little green oranges are, as big as half the end of my little finger. Here were also a great variety of other exotique plants and several labyrinths and a pretty aviary."

Some ten years before this incident, in 1654, we have Robert Sharrock, the Winchester cleric and botanist, asking how ever any gardener could preserve anything at all in their "Houses of Defence." "It is to be wondered," he said, "how the gardeners get delicate plants to live by sheltering them in dark places during the winter. Some defend the myrtle, pomegranate and such other tender plants whether by houses made of straw like beehives, or of boards (with inlets for the sun by casements or without them) litter of horses' stable being laid in very cold weather about the house of defence." In his *History of the Propagation and Improvement of Vegetables*, printed at Oxford in 1660, he describes hot-beds of dung covered with glass casements and he shows by his further descriptions that he knew light and air were just as important as heat, although neither he nor his gardening colleagues quite knew how to effect the improvement. For in considering the Reverend Sharrock's observations it is well to remember that before the introduction of the orange in 1562 the need for the preservation of plant life in winter-time had been negligible. But the number of casualties with the orange must have been both great and disheartening to those early gardeners, before a reasonably constructed orangery with a comparatively efficient heating apparatus had been devised. After all, the first free-standing iron stove was not made in Holland until 1670, according to horti-cultural writers of the time, so that heating before this could only have been in the nature of open hearths, brazier types of fire, or charcoal pans, not forgetting a few thick candles to attemper the air.

Sir Hugh Plat, mentioned previously for his experiments with heating, was among the first to advocate the desirability of all noblemen and gentlemen of taste having a garden house or covered garden.

"I hold it one of the most delicate and pleasing things to have a fair gallery, or great chamber or other lodging that opens fully upon the east or west to be inwardly garnished with sweet herbs and flowers, yes and fruit if it were possible. For the performance thereof I have thought of these courses, following, for you may have fair sweet marjoram, basil, carnation and rosemary to stand loose upon shelves, here to enjoy the warm sun or temperate rain at your pleasure. In every window you make square frames either of lead or boards well pitched within, fill them with some rich earth and plant such flowers or herbs therein as you like best [the first window box gardening?].

"In the shady parts of the room you may prove if such shady plants as do grow abroad out of the sun will not also grow there as sweet briar, bays, germander, but you must often set open your casements especially in the day time. You may also hang about the sides of the room small pompion or cucumber pricked full of barley."

Of Dutch rose or carnation growing in the winter Sir Hugh says: "Place them in a room that may in some way be kept warm either with a dry fire or with the steam of hot water conveyed by a pipe fastened to the cover of a pot that is kept seething over some idle fire, now and then exposing them in warm days from twelve to two in the sun or to the rain if it happen to rain, or if it rain not in convenient time set up pots having holes in the bottom in pans of rainwater and so moisten the roots."

He goes on to advocate a house built especially for the forcing of fruit. Traditionally the roof was opaque, but Sir Hugh knew the value of light and airiness, a lesson it took the gardeners who followed him many more years to learn and apply. "If we are to have early fruit and do not regard labour a charge," he said, "then let us build a square and close room having many degrees of shelves one above the other in which we may aptly place so many of these dwarf trees as we shall think good; in time of cold weather we may keep the same warm in nature of a stove with a small fire being made in such furnace. . . .

"If the weather be fair and open and the room be made full of windows or open sides we may for such time use the benefit of sunshine or carry them abroad at our pleasure. . . ."

He further suggested that the room could be heated with steam from a vessel where the beef was boiling and then in the room, he told his readers, they would be able to grow oranges, lemons, pomegranates, coloquintida and pepper-trees. The sides of the room might be plastered and the top covered with canvas as a removable ceiling.

He then asked if it would be a good thing to let in pipes of lead to breathe steam out at the end only, "or else at divers small vents which may be made in that part of the pipe which passeth along the stove."

Plat wondered if that idea were a mere conceit because the steam of water, he added, would not extend far, "but if the copper be of metal and made to close so that no air can breathe out save to the pipe which is sodered or well closed in some part of the cover, then it seemeth probable because the cover may be put on after the pot has simmered."

This was thinking much too far in advance of the times and for many, many years such matters were forgotten by the horticultural builders, for when one of the earliest and most pretentious of early plant conserves was built at the Oxford Physic Garden, the stone being laid in 1621, no heating was incorporated in the building whatsoever. Yet it was a good building for its time, as the drawing of the garden in a plate of 1675 after Loggan shows. In this, the first conservatory or shelter for "tender greens," not for oranges, was shown as a long, low stone-built orangery type of structure with arched windows and solid slate roof, more like a northern Methodist Chapel than a greenhouse. It was 60 feet long and was of such solid construction that early in the eighteenth century it was transformed, by adding a storey and other alterations, into a herbarium, library and professor's house.

It was true that the massive stonework open to the south would pick up sun heat during the day and radiate some of that heat during the night, and the same solidarity of structure would also serve to keep out the cold. If it became very cold there was a

c

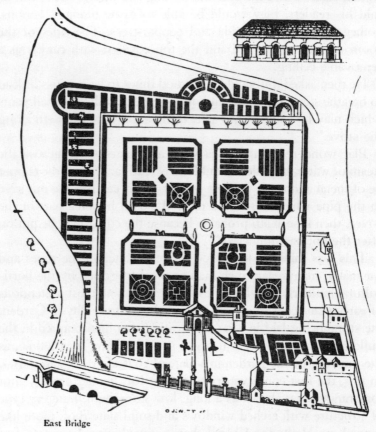

An early plan of the Oxford Physic Garden showing an elevation of
the old "Conservatory for tender greens."
(*Oxford Botanic Garden*, R. W. T. Gunther, 1912.)

most ingenious method brought into use to warm the tender greens. A four-wheeled wagon, looking like a watchman's oblong brazier on wheels, was filled with charcoal, set alight, and then drawn backwards and forwards through the conservatory by the gardener.

Jacob Bobart, the elder, a native of Brunswick, was the first gardener appointed at the Oxford Physic Garden and in a catalogue of plants grown in the garden and published in 1648, of the 2,000 described only some 600 were natives; showing in some measure the need for artificial protection in the garden.

It is to John Evelyn that we are again indebted for some of the earliest detailed accounts of covered and indoor gardening of the time. He was, wherever he travelled, and he travelled much, always looking out for new ideas and new techniques in gardening, and always had a particularly keen eye for the newer art of his day, the art of the conservatory and greenhouse.

In 1677 Evelyn was at Lord Arlington's home at Euston in Suffolk and, writing of the orange garden, said it was a fine one and led into the greenhouses at the end of which was a hall to eat in and the conservatory, some 100 feet long, "adorned with maps, as the other side is with the heads of Caesars ill cut in alabaster; overhead are several apartments."

At Chelsea in 1680 he noticed "the curiously contrived conservatory" recently built that year at a cost of £138. "What was very ingenious was the subterranean heat conveyed by a stove under the conservatory which was all vaulted with brick so as he [Mr. Watts, keeper of the Garden for the Apothecaries Company] has the doors and windows open in the hardest frosts, secluding only the snow." One wonders just how much the open air treatment was due to choice or because of the deadly fumes from the ingenious arrangements or the considerable body of uncontrolled heat generated which might well have burned up the plants but for the open window treatment.

There is an excellent engraving in a Dutch publication of 1670 of an orangery of the day, one of the few plates of an interior to be found, and in it can be seen the heavy wooden-beamed flat roof, windows only on the one side and these latticed with large masonry pillars between them. On the back wall are murals in bas-relief and pictured are a lady and gentleman in the costume of the times of Charles II, strolling nonchalantly among the orange-trees in tubs on the floor, while two gardeners struggle through the far door with a tub carried on a kind of stretcher arrangement. Below the murals on the back wall can be seen two free-standing tile-fronted Dutch stoves with their stacks leading out to the roof. It is obvious that all stoking and cleaning of the stoves would have to be done inside the building.

Evelyn in his travels on the Continent must have seen this type of building and taken a keen interest in both this and other Continental methods of heating by wall flues, stoves, and earthen-ware pipes, the latter the first means of conveying heat from stoves to be used by the Dutch. As a result of this John Evelyn invented his own stove for a hot-house which proved both satisfactory and reasonably efficient. We first hear of this newly invented stove in Evelyn's *Kalendarium Hortense*, published in 1664, where the author notes that plants preserved in stoves or greenhouses heated with iron stoves, charcoal, or "subterranean caliduits" while they managed to live through the winter did not do so without spoiling "their complexions" by way of leaf parch-ing, drying, yellowing and falling. The reason for this, surmised Evelyn, was because the plants were deprived of fresh untainted air and his invention was designed to remedy that.

His idea was to push out the "imprisoned and effete air" with a constant stream of fresh and untainted air. The greenhouse he had in mind was one of which the depth should not exceed 12 or 13 feet nor the height above 10 or 11 at the most and "it being placed at the most advantageous exposure to the sun, that side be made to open with large and ample windows or chasses (for light itself next to air, is of wonderful importance) the joints and glazing accurately fitted and cemented." To keep out a rush of cold, crude air he advocated a small porch of which the door might be shut before the greenhouse door was opened.

"At one of the ends of the conservatory or greenhouse," his

instructions continued, "(tis not material whether east or west) erect on the outside wall your stove be it of brick, or (which I prefer) of Rygate stone, built square of the ordinary size of a plain single furnace (such as chymists use in their laboratories for common operations) consisting of a fire hearth and an ash hole only; which need not take up above two feet from out to out: let it be yet so built that the fire grate stand about three feet higher than the floor or area of the house."

For the conveyance of the heat from his oven Evelyn recommends three or four pipes of crucible earth, stating that sea coal or charcoal would melt iron ones. These pipes were to go through the fire with their "noses" projecting at both sides so that fresh warmed air would flow through the pipes and circulate freely in the greenhouse.

At the end of the house was a hole communicating with an under-floor pipe which continued back to the stove up the side of it and into the ash-hole just underneath the grate. The drawing power of the fire was then to draw the stale air of the conservatory into this flue and up the chimney, making room for new heated air coming in to balance it.

Evelyn goes on to give details of what to put in the house for "hybernising" and says:

"September: About Michaelmas (sooner or later as the season directs) the weather fair and by no means foggie, retire your choice greens and rarest plants (being dry) as orange, lemons, Indian and Spanish jasmine, Oleanders, Barba-Jovis, Amomum plin., Citysus lunatus, Chamalaca tricoccos, Cistus ledon clusii, dates, aloes, sedums, etc into your conservatory, ordering them with fresh earth . . . but as yet leaving the doors and windows open, and giving them free air so the winds be not sharp and high nor weather foggie, do thus till the cold being more intense, advertise you to inclose them altogether."

For plants which would not endure the conditions of the house he advocated that "the pots be placed two or three inches lower than the surface of some bed under a southern exposure and then covered with glass. Thus you shall preserve Marum syriacum,

Cistus, Geranium noste olens, Flos cardinalis, Maracoco, Arbutus Ranunculus, Anemonies, Acacia aegypt, etc."

For November his instructions were to "quite enclose your tender plants and perennial greens, shrubs, etc., in your conservatory, secluding all entrance to cold and especially sharp winds, and if the plants become exceedingly dry, and that it do not actually freeze, refresh them sparingly with qualified water mingled with a little sheep or cow dung; if the season prove exceeding piercing (which you may know by the freezing of a dish of water or moistened cloth set for that purpose in your greenhouse) kindle some charcoals, and when they have done smoking put them in a hole sunk a little into the floor about the middle of it; unless your greenhouse have a subterranean stove which moderately and with judgement tempered is much to be preferred."

He went on to recommend that on all good sunny days without "the least wind stirring" then the correct treatment for plants was to "shew them the light through the glass windows (for light is half their nourishment philosophically considered)."

If the weather was very cold and the gardener had to keep the house enclosed and mustiness occurred, then he had to make a fire in the stove and open all windows from ten till three, "closing the double shuts (or chasses rather) later," whilst he continued with a gentle heat not renewing the fire until nightfall.

In 1694 Sir Dudley Cullum, of Hansted in Suffolk, had actually made and used Evelyn's stove and thanked that gentleman for it. "Sir," he wrote, "I cannot but think myself obliged in gratitude to give you an account how well your lately invented stove for a greenhouse succeeds (by the experience I have had of it) which certainly has more perfection than ever yet art was before master of . . . by this free and generous communication of yours you must have highly obliged all the lovers of this recreation. . . ."

In *The Art of Gardening* by Evelyn, with notes by the Reverend Mr. Lawrence, there is reference to three types of house, "the conservatory, the common greenhouse and another sort of conservatory."

The conservatory and common greenhouse sound uncommonly alike. The common type has a floor higher than the surrounding ground, with a stove in a vault underneath in which the fires were kept ready laid for lighting when the water froze "upstairs."

The "other sort" is obviously the poor, or comparatively poor, man's greenhouse. Probably the younger sons of the nobility had to make do with such "an easily erected" structure, full details of which are given, starting with the marking out of a piece of ground in autumn under a warm, south wall of about 12 or 15 feet in length and about five or six feet in depth. On this ground the gardener was to prepare a wooden frame of some seven feet high at the wall side, lowering to about five feet at the front. The framework was then to be filled with dry reeds or rushes, with a door of rushes. The top "or the greater part thereof" was to be covered "with common glass frames like those of hot-beds, about three of them in the whole length of the front."

It is an interesting comment on the times and its lack of variety in plants to preserve when Evelyn goes on to say such a conservatory would normally hold 25 greens in pots besides flowers and other small plants.

It must have been in structures similar to one of these houses, or maybe in a mixture of all three, that Bishop Henry Compton, of Fulham Palace, raised so many of his "exotics" during the reign of James II, some time about 1688 when his gardens and greenhouses were said to contain a greater collection of plants than any in England, although the term "exotics" was surely a high-sounding one for the many half-hardy economic and medical plants which must have comprised his collection as well as "tender greens."

But Bishop Compton was certainly one of the first to encourage the importation and raising of ornamental exotics and "was very curious in collecting them as well as in cultivating." In his stoves and greenhouses he was said to have about 1,000 species of exotic plants, and many "that had been previously esteemed too tender to expose unprotected in our climate."

Quite exhaustive details for conserving were given by Henry van Oosten in *The Dutch Gardener* of 1703 quaintly dedicated: "The design of this book is to serve the young lovers of flowers." Many lessons had already been learned on the Continent and here, sound advice was being given on over-wintering the members of the citrus family, which after all were still the principal concern of the English gardeners of the time. It was apparent that in Holland, at least, much had been learned of the value of light, air and sunshine, and the killing power for plants of a damp, moisture-saturated atmosphere and the dangers of unregulated, badly placed heating was also well understood.

Van Oosten told his English readers the secrets of the Dutch growers, and for those who could, and did mark and learn, there was progress to be made, although there had obviously been serious trouble with the Dutch heating methods. Van Oosten fulminated against all stoves or heating of greenhouses whatsoever. He warned his readers of the dangers while at the same time enthusing on the delights of orange growing. "I affirm upon very good grounds," he said, "that in all the compass of gardening there is not a plant or tree that affords such extensive and lasting pleasure, for there is not a day in the year on which the orange trees may not and indeed ought not to afford matters of delight whether it be in the greenness of their pretty leaves or in the agreeableness of their form and figure or in the pleasant smell of their flowers or in the prettiness and (duration). . . .

"The stove or winter house ought to be handsomely contrived to the end that the master or proprietor when he sits in his pleasure house may not only entertain himself with the sight of the trees but enjoy their agreeable smell through open windows.

"Orange trees are inconvenienced by the fires either great or small whereof most make use of in their stoves and greenhouses; if small the heat can work but on them that stand near it without touching that at least which is farther off. If you make it below and in a few places as is commonly done then the heat cannot work nigh on the tops nor on the side where the fire is not and if you make the fire in a higher place then it cannot reach the lower-

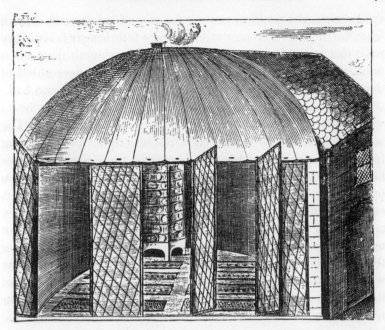

An early Continental type greenhouse with boarded and tiled roof, but lattice windows from floor to ceiling giving full light to the south. Plants, as will be seen, are grown in the ground in beds and a giant tiled hot-air stove is in the background.

(*Neue Garten-Lust*, H. Hesse, Leipzig, 1714.)

most twigs. On the contrary if the fire be large it will without doubt dry up the outward bark of the trees and twigs.

"Therefore banish all fire from out of the house wherein we winter our orange trees.

"To make a good wintering place there are five qualities:

1. That it be well placed against the sun.
2. Well provided with windows which may shut close if there be occasion.
3. It must be thick and well provided with firm walls.
4. Well ceilinged.
5. Not hollow under the floor.

"The best position is against the south so that the sun may shine into the house from 9 or 10 in the forenoon till it sets in the evening, the position against the east where the sun shineth from the time of its rising until noon or a little longer is also very well, that towards the west wherein the sun shineth from noon until it sets may be made use of when you cannot have either of the two first. That against the north is very dangerous and bad.

"Concerning the airiness. It is required that the doors be so wide that orange trees may be easily carried in and out. The windows must be large and high, reaching quite to the ceiling from the breastwork which is commonly three feet high. The breadth of the windows must be five or six feet, that when you open them in the winter when the sun shineth brightly, as it is necessary, the sun may shine on them all at once whereby they are mightily refreshed and the least moisture that may be within may be dried up by its rays. These windows must have another frame within of oiled paper and one without of glass, for the wooden shutters are of no moment and cheat many lovers of flowers. These frames must be very closely fitted in the winter time that the air may not penetrate into the house through the least hole for this is powerful enough to alter the temperate air

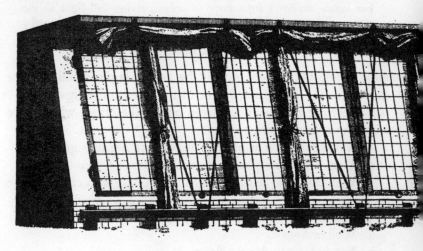

An example of a Dutch forcing-house of 1737—still with use of glass. Note the neat arrangement for shading. En
(*Genleggen van pragtige en gemeene lan*

that would remain in the greenhouse even since the last fair day without which the orange trees cannot retain its firmness. The room over it must be very close for the cold and moisture can penetrate through the roof as well as through the side, whereof the floor thereof or the boards must be deck laid in, and on it, if it be not inhabited in winter, be laid hay or straw and the windows must be closely shut.

"Concerning the material of the floor they may be made of hard ground or plaster or of boards, and these are the best. Above all things you must take care that underneath it be no vault or cellars for these are mortal to all orange trees, lemons, jasmines, and other trees planted in boxes because these low and hollow places are commonly moist and unfrequented by the beams of the sun without which the winter house cannot be well constituted.

"Concerning the length and breadth of the greenhouse they may be 24 or 36 feet more or less if they be fair and dry, so that neither cold nor moisture can penetrate, for it is not the beams of sun that shine immediately on the leaves of the orange tree that are essentially wholesome to them because they seldom touch

and still lean-to in style, but with under-floor heating and good
by way of the front glass hinged like a door.
r de la Court, Leyden, 1737.)

the inward leaves of the crown, but those that shoot into the hollowness of the greenhouse hinders that there be no moisture left to do hurt.

"If the water in the greenhouse is frozen then you must gently warm the trees or rather the leaves with burning lamps so hung that the flames thereof may not touch the trees." The trees, he advised, should not be taken out of the greenhouse until the full moon of April was past.

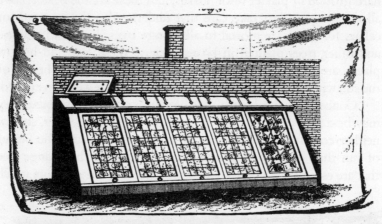

A Dutch forcing-house of 1737 with back wall flue, solid ends, entrance through the front glasses, but with a minimum of opaque (in this example, wooden ventilator) roofing.

(*Genleggen van pragtige en gemeene landhuizen,*
Pieter de la Court, Leyden, 1737.)

It is very noticeable during this period that the terms green-house and conservatory were used synonymously and noticeable too that both words sprang from the actual gardener's vocabulary, for after all the greenhouse was just that, a house in which to preserve tender greens, while the conservatory was a conserve for plants. It should not be lost sight of, either, that apart from the orange the gardener's plants were green, for there was little colour or blossom to commend them; candidates for over-wintering at this time being the citrus family, myrtles, pomegranates, the bays, the succulents and the cypress.

If the seventeenth-century gardeners had been fortunate enough to have, and had been able to conserve, but one-half of the magnificent floral treasures of later centuries we might well have been talking about flower-houses now instead of green-houses. It was only with the greater variety of plants available in the nineteenth century and the different means available to conserve them that the conservatory, stove and greenhouse came to have their own specific meanings.

3

A CONDUCTED TOUR

To return to the early days of the greenhouse art, we have a most delightful and unusually detailed description of the transitionary period between hot-houses and stoves for tender greens and fruit and greenhouses for flowers given in an old manuscript which the Reverend Dr. Hamilton presented to the Society of Antiquaries in 1794.

It is nothing more or less than a personally conducted tour in 1691 by one Mr. J. Gibson of "several gardens near London, with remarks on some particular wherein they excel or are deficient."

The fact that Mr. Gibson made his visits in January says something for his perspicacity, for if a greenhouse looks well at that time then the gardener in charge is a skilled one. We should say that of such a visit today, yet it is probably much nearer the truth to say that if Mr. Gibson had visited the same greenhouses in high summer he would have found them devoid of anything but the staging. Their winter inmates would have been outside gracing the parterres and knots or lining the many walks and mazes of the gardens of the time.

One could have wished for a little less vague term than that of "greens" with which Mr. Gibson fills so many houses, but I think it would be fair to say when this painstaking visitor says "greens" without specifying any particular plant or plants, he can be assumed to be referring to the "greens" that were the usual and common types in collections of his day, such as succulents, cacti, the orange- and lemon-trees, laurustinus, sassafras, phyllyrea, myrtles, oleanders, guavas, papaws, agaves, aloes, spurges, acacias, cannas, figs of various kinds, arums, bulbs—hyacinth and tulip—the passion flower, with some such economic plants as coffee, ginger, arrowroot and bread fruit. It would be true to say

4

that Mr. Gibson would see very little colour either of flower or foliage to reward his pilgrimage; all that had still to come.

It would no doubt be most interesting for us to have had fully detailed plant lists for each establishment, yet I think we should be more than satisfied, and not a little surprised, to know that there were so many prosperous gardens and so many fine greenhouses and stoves to see in January 1691.

But let Mr. Gibson start on his tour. First he goes to that still-popular resort of the tourist and tripper of today—Hampton Court.

"Hampton Court garden is a large plat, environed with an iron epalisade round about next the park, laid all in walks, grass, plats and borders. Next to the house, some flat and broad beds are set with narrow rows of dwarf box, in figures like lace-patterns. In one of the lesser gardens is a large greenhouse divided into several rooms, and all of them with stoves under them, and fire to keep a continual heat. In these there are no orange or lemon trees, or myrtles, or any greens, but such tender ones that need continual warmth."

Back to the city and at Kensington Gardens the orange, lemon, myrtles and what other trees they had there in summer, were all removed to Mr. London's and Mr. Wise's greenhouse at Brompton Park, a little over a mile from the Gardens.

To the suburbs again, where the Queen Dowager's garden at Hammersmith had a good greenhouse, "with a high erected front to the south, whence the roof falls backwards. The house is well stored with greens of common kinds; but the Queen not being for curious plants and flowers, they want of the most curious sorts of greens, and in the garden there is little of value but wall trees; though the gardener there, Monsieur Hermon Van Guine, is a man of great skill and industry, having raised great numbers of orange and lemon trees by inoculation, with myrtles, Roman bayes, and other greens of pretty shapes which he has to dispose of."

The orangery to beat all orangeries was obviously the one at Beddington Garden, then in the possession of the Duke of

Norfolk "but belonging to the family of Carew," mentioned earlier.

"It has in it the best orangery in England. The orange and lemon trees there grow in the ground, and have done so near one hundred years, as the gardener, an aged man, said he believed. There are a great number of them, the house wherein they are being above two hundred feet long; they are most of them thirteen feet high and very full of fruit, the gardener not having taken of so many flowers this last summer as usually others do. He said he gathered off them at least ten thousand oranges this last year. The heir of the family being but five years old, the trustees take care of the orangery, and this year they built a new house over them. There are some myrtles growing among them but they look not well for want of trimming. The rest of the garden is all out of order, the orangery being the gardener's chief care. . . ."

By the river the old-established Chelsea Physic Garden is next on the itinerary and here the garden "has great variety of plants, both in and out of greenhouses, their perennial green hedges and rows of different coloured herbs are very pretty, and so are their banks set with shapes of herbs in the Irish stitch-wat, but many plants of the garden were not in so good order as might be expected."

The Earl of Devonshire's place, Arlington Garden, gets a black mark from Mr. Gibson: "Their greenhouse is very well, and their greenyard excells; but their greens were not so bright, as if they suffered something from the smutty air of the town.

"My Lord Fauconbergh's Garden at Sutton Court, has several pleasant walks and apartments in it," reports Mr. Gibson next. The upper garden he thinks is too irregular and he does not like the bowling-green. However, "the greenhouse is very well made, but ill set. It is divided into three rooms, and very well furnished with good greens; but it is so placed that the sun shines not on the plants in winter, where they most need its beams, the dwelling house standing betwixt the sun and it."

He kept "Italian bayes" inside during the winter for putting out in cases during the summer, to enclose the apartment for white pheasants and partridges.

Sir William Temple had lately gone to live at Farnham so Mr. Gibson reports: "His garden and greenhouse at West Sheen, where he has lived of late years, are not so well kept as they have been, many of his orange trees and other greens being given to Sir John Temple, his brother at East Sheen, and other gentlemen; but his greens that are remaining (being as good a stock as most greenhouses have) are very fresh and thriving, the room they stand in suiting well with them and being contrived, if it be no defect in it that the floor is a foot at least within the ground, as is also the floor of the dwelling house." He had attempted to grow orange-trees in the ground (as at Beddington) and for that purpose had enclosed a square of ten feet wide with a low brick wall and sheltered them with wood, but they would not do. "His orange trees in summer stand not in any particular square or enclosure under some shelter, as most others do, but are disposed on pedestals of Portland stone at equal distance, on a board over against a south wall, where is his fairest fruit and fairest walk."

In Sir Henry Capel's garden at Kew in summer "stood out his orange trees and other choicer greens in two walks about 14 feet wide, enclosed with a timber frame about seven feet high and set with silver firs hedge-wise which are as high as the frame, and this to secure them from wind and tempest and sometimes from the scorching sun."

At Chiswick, not far away, at Sir Stephen Fox's place the greenhouse was well built, well set and well furnished. He had two myrtle hedges, which he did not take in, for they were planted in the ground, but he covered them in winter with painted board cases.

Disaster had overtaken Sir Thomas Cooke at Hackney, when Mr. Gibson called, for "There were two greenhouses the greens in which were not extraordinary, for one of the roofs had been made a receptacle for water and, overcharged with weight, had fallen down the year before upon the greens making a great destruction among the trees and pots."

Sir Robert Clayton at Morden in Surrey got full marks for his

D

big plantations, for with soil not "very benign to plants, but with great charge he forces Nature to obey him.

"He built a good greenhouse," says the discerning Mr. Gibson, "but set it so that the hill in winter keep the sun from it, so that they place their greens in a house on higher ground not built for that purpose.

"The Archbishop of Canterbury's garden at Lambeth has little in it but walks, the late Archbishop not delighting in gardens, but they are now making them better; and they have already made a greenhouse, one of the first and costliest about the town. It is of three rooms, the middle having a stove under it; the foresides of the rooms are almost all glass, the roof covered with lead, the whole part (to adorn the building) rising gravel-wise higher than the rest; but it is placed so near Lambeth Church, that the sun shines most on it in winter after eleven-o'clock, a fault owned by the gardener"—one can imagine him with a sly grin asking Mr. Gibson what could be expected of these church dignitaries who thought they were gardeners.

The fault was laid at the door of "the contrivers." Most of the greens were oranges and lemons with very large fruits on them.

At Mr. Evelyn's "pleasant little villa at Deptford" was a fine garden for walks and hedges (especially his holly one which he writes of in his *Sylva*), and a pretty little greenhouse, with an "indifferent stock in it."

A new house and garden of Mr. Watt, near Enfield, showed that "He built a greenhouse this summer with three rooms (somewhat like the Archbishop of Canterbury's) the middle with a stove under it and a sky light above, and both of them with glass on the foreside, with shutters within, and the roof finely covered with Irish slate. But this fine house is under the same great fault as that of the Archbishop's, Lord Fauconbergh's and Sir Robert Clayton's, they were built in summer, and thought not of winter; the dwelling house on the South side interposing betwixt the sun and it, now when its beams should refresh the plants."

Enfield, where the great Dr. Uvedale had his garden, was also favoured by Mr. Gibson where our horticultural tourist says "the

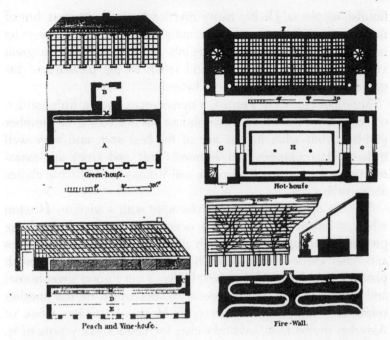

Green-houſe.

Hot-houſe

Peach and Vine-houſe.

Fire-Wall.

A typical greenhouse (synonymous with orangery) of the day, with flued walls and slated roof. The hot-house is again a very solid structure with tan-pit and flued walls. The peach- and vine-house is good practice for the day with long sloping roof and flued back wall.

(*The Gardener's Daily Assistant*, John Abercrombie, 1794.)

garden at Enfield is observable for nothing but his greenhouse, which he has had for many years. His orange, lemon and myrtle trees are as full and furnished as any in cases. He has a myrtle tree cut in the shape of a chaire, that is at least six feet from the case, but the lower part is thin of leaves."

At East Barnet, where Mr. Richardson had a pretty garden, but also had the gout, they had orange- and lemon-trees. The wife and son being managers of the garden, says Mr. Gibson, because the husband's gout disposed him to not managing it himself "they cannot prevail for a house for them (the oranges and lemons) other than a barn end.

"Captain Forster's greenhouse at Lambeth is full of fresh and

flourishing plants. He has many myrtles, both the greatest, but of
the most fanciful shapes, that are anywhere else. . . . Of flowers he
has a good choice, and his Virginia and other birds in a great
variety, with his glass hive, add much to the pleasure of his
garden." A delightful prospect indeed.

Monsieur Anthony Vesprit, a nurseryman, had "a little garden
of very choice things. His greenhouse has not very great number
of plants, but what he has are of the best sort, and very well
ordered. His oranges and lemons (fruit and tree) are extraordinary fair, and for lentiscus's and Roman bayes he has choice
above others."

We are nearing the end of the tour with a visit to Hoxton
where a call at Mr. Ricketts's nursery shows he "has a large
ground abundantly stocked with all manner of flowers, fruit trees
and other garden plants with lime trees, which are now much
planted; and, for a sale garden, he has a very good greenhouse,
and well filled with fresh greens, besides which he has another
room very full of greens in pots. He has a greater stock of
Assyrian thyme than anybody else; for besides many pots of it,
he has beds abroad, with plenty of roots, which they cover with
mats and straw in winter. . . ."

Almost adjoining is Mr. Pearson, who had a great choice of
flowers, anemones he claimed to be the best in London which
he only sold to gentlemen. "He has no greenhouse, yet has
abundance of myrtles and striped philareas, with oranges and
other greens, which he keeps safe enough under sheds, sunk a foot
within the ground, and covered with straw. He has abundance of
cypresses which, at three feet high, he sells for fourpence apiece
to those who take any number. He is moderate in his prices and
accounted very honest in his dealings which gets him much
chapmancy."

Evidently a good nurseryman to know.

4

A HOT-HOUSE COTERIE

ONCE we get to the eighteenth century, there is ample evidence of artificial heat, hot-houses and greenhouses, from all parts of Britain. A revealing glimpse into a coterie of the eighteenth-century botanists, scientists, plant-lovers and gentlemen, all drawn together by the love of plants, flowers and experiment, is contained in the letters of Richard Richardson, of Bierley, Bradford, Yorkshire, who was a medical man seemingly with plenty of time to give to his hobby—plant-rearing in his stoves and writing to most of the other prominent plant-lovers and botanists of his day, both in this country and abroad.

In the letters appear all, it would be fair to say, the practising gardeners (as distinct from pure botanists and lovers of the curious) so many of whom were part of the eighteenth-century background; all, without exception, showing in their letters a keen and knowledgeable interest in the subject.

In the North of England John Blackburn, of Orford near Warrington, and Dr. Richardson himself are stated to have erected the first hot-houses.

There is a letter to the Doctor from Dr. Uvedale, writing in 1696 of garden losses during a particularly hard winter—losses caused by cold and frost—where he speaks of his stoves having suffered owing to the illness of a gardener, whose instructions had not been carried out properly in keeping up the necessary temperature or allowing the necessary airings. At that time Dr. Uvedale was receiving seeds from friends at the Cape and in Rome.

The Sherards, W. & J., who were both eminent botanists, in a letter to their Bierley correspondent evidently knew quite a lot about stove and greenhouse practice. They had learned from experience and were happy to pass on their knowledge to Dr.

Richardson who had just completed building his stove in 1718. They had obviously been over to Leyden where they had made an expert and designing survey of the stove built in the famous botanic gardens there by the celebrated physician and botanist Dr. Boerhaave.

"The secret of the stove," the Sherards told the Doctor, "consists in making the angle of the shutters equal to the angle of the Pole which with you is about 52 degrees. This causes the sun to fall in a straight line without any angle of reflection. The other secret, in philosophy, is the having of the glass of the stove to go to the top of it that there may be no place left at the top where the sun does not shine on it. If any such place be left in the shade the vapours will be raised into it and after the sun is off, fall down on to the plants and mould them, the making of the back part of the stove sloping, and glass to the top prevents this."

This was excellent advice for the time, but it was advice and experience which took a long time to become standard practice, and referred only to the lean-to house. It was only people like the Richardsons, the Sherards, the Sloanes and the Petres of the age, the gentlemen-scientists, in which the century abounded and who were its glory and its pride, who either knew of or were in receipt of such sound practical advice. These were the Royal Society gentlemen, whose enthusiasm for botany spilled over into the fields of mathematics, meteorology, antiquarianism, experimental philosophy and invention.

Lord Petre, eminent florist with a beautiful garden and establishment at Thorndon in Essex, writing in 1734 to Bradford, is able to say, with justifiable pride, that he has, he believes, "the greatest part of stove plants, whether succulent or others, as are yet known in England."

Peter Collinson, of whom more later, the collector, naturalist and botanist with a botanical garden at Mill Hill, was a friend and collector for Lord Petre, and after Lord Petre's death in August, 1742, he was bewailing the fact that no one but Mr. James Gordon, of Mile End nurseries, was left with a good stove-house and that "all stove plants would go down."

Fortunately his pessimism was not well founded and, writing from Thorndon in April, 1746, when he was a guest of Lady Petre, he gives a delightful picture of a glasshouse range of the time and took the trouble to catalogue the main inhabitants and to describe them lovingly. Through his eyes we can see the houses now.

There was not such a collection in all England, he said, except for Oxford or Chelsea, but there were many more plants at Thorndon than even those two exotic plant establishments possessed.

"The great stove is the most extraordinary sight in the world," he told Dr. Richardson. "All the plants are of such magnificence and the novelty of their appearances strikes everyone with pleasure. The trellises all around are covered with a species of passion flower which run up near to 30 feet high, the creeping, great-flowering cereus (torch thistle) blows annually with such quantities of flower that surprises everyone with their beauty and, at the same time, perfume the house with their scent.

"There is a variety of cereus that have carried up their perpendicular heads to the very top; papaws (*Carica papaya*) both male and female; guavas (*psidium*), plantain (*bananas*), several sorts of palms, dragon trees (*dracaena*), *ficus malabharicus*, *Rosa chinensis*, acacias, bamboos; besides a great variety of species received from New-Spain, seeds that I don't know. Many have flowered and fruited; the papaws being now full of fruit and blossom; and then there is a most delightful show of several sorts of white lilio-narcissus now in flower.

"The lesser stove, 60 feet long and 20 feet wide, is full of a vast variety of all species of tender exotics."

Quite a father-figure and a rare professional among the many amateurs of his time, with a down-to-earth practicality, was Philip Miller, F.R.S., gardener to the Worshipful Company of Apothecaries at their Botanical Gardens in Chelsea. In Miller's celebrated *Dictionary*, first published in 1731, he has many sound things to say about greenhouses or conservatories, conservatories for conserving, mind you, not orangeries, and not greenhouses for adding novelty and gardenage to your living quarters. The furtherance of botanical and herbal studies, as in Miller's case, was still uppermost in all minds.

The era had not yet arrived of horticulture for beauty's sake; this was still an age of study and classification, and so Philip Miller in his writings upon the greenhouses as he knew them in the early part of the eighteenth century is thinking all the time of plants for medical use or for study by the many amateur and professional botanists of his day.

His greenhouse, for which he gives such admirably detailed instructions, is far away from being like the houses that later housed with success the floral wealth of almost the whole world. It was very much an architectural piece of work, not that of a horticultural designer. But Philip Miller was a most successful raiser of plants in his day, and one of the first to find that he could raise difficult plants by placing the seeds in a bed of tan. He, too, like all good gardeners, was never too old to learn, and in a later edition of his *Dictionary* he amended the plans of his greenhouses to show a glass roof.

His first house was a ceilinged one at Chelsea, but even with this he must have had many successes with rare plants or the Chelsea Physic Garden would never have become, as it did, the Mecca of plant-lovers and lovers of the curious to see what was new with Miller, so that it became a must on the itinerary of all botanists visiting London, not only from this country but from abroad.

The greenhouse at Chelsea was first erected in 1680 at a cost of £138, and a further greenhouse and two hot-houses were built and finished for Miller in 1732. It was from his "apartment in the greenhouse" there that he compiled his *Dictionary*, where, under the heading "Greenhouses and Conservatories," he says; "As of late years there have been great quantities of curious exotic plants introduced into the English garden so the number of greenhouses or conservatories has increased, and not only a greater skill in the management and ordering of these plants has increased therewith, but also a greater knowledge of the structure and contrivance of these places so as to render them both useful and ornamental; since there are many parts to be observed in the construction of these houses whereby they will be greatly improved, I thought it necessary not only to give the best instructions for this I was capable of, but also to give a design of one in the manner I would choose to erect."

The most practical Mr. Miller then goes on to give the soundest advice to would-be owners of such edifices, length, he said, depending on numbers of plants or fancy of the owner, but depth never greater than the height in the clear, which in small or middling houses may be 16 or 18 feet but for large ones from 20 to 24 feet is a good proportion, "for if the greenhouse is long and too narrow it will have a bad appearance both within and without nor will it contain so many plants if proper room be allowed for passing in front and on the backside of the stands on which the plants are placed; and on the other hand if the depth of the greenhouse is more than 24 feet there must be more rows of plants to fill the house than can with conveniency be reached in watering and cleaning, nor are houses of too great depth so proper for keeping of plants of moderate size.

"The windows in front should extend from about $1\frac{1}{2}$ feet above the pavement to within the same distance of the ceiling which will admit a cornice round the building over the heads of the windows."

The house, he said, should be of stone or brick then it would be possible to have a house over the greenhouse. At the back of it should be a tool-house, thus stopping the frost entering that way.

Miller advocates that his floor, paved or tiled, some two or three feet above the earth, should preferably be arched to stop the damp rising in winter, which was so harmful to plants. And when the air was often too cold to be admitted to the house, he advised, as did everyone at the time, and as was the case for nearly another one hundred years, that a flue should be built under the floor about two feet from the front of the house, about ten inches in width and two feet deep, to be carried the whole length of the house, to be returned along the back wall and "carried up in proper funnels adjoining the tool house by which the smoke may pass off." The fireplace was to be at one end and contrived so that it could be fired from the tool-house.

It was obvious that flues had been responsible for many dismal failures, for the poisoning and blackening of rare plants brought from afar and raised with infinite patience only to die by suffocation or drowned in soot, and Miller finds himself having to apologise for having mentioned them at all. He points out that many people will be surprised to see him advocating "flues of ill consequence as indeed they have often proved when under the direction of unskilful managers who have thought it necessary whenever the weather was cold to make fires therein, but however injurious such flues have been under such management when skilfully managed they are of very great service."

He advocated stout shutters for inside the windows at least $1\frac{1}{2}$ inches thick, which if made close enough, would keep out common frost and these had to be hinged for he had seen, he says, if people had not made fires and had not made closely fitting stout shutters, people had to stuff their shutters with straw and their windows with mats and then, if the frost continued for a day or two, the same windows be kept closed it had "proved very injurious to the plants." He advocated opening up even for the two or three hours of sunshine which often was enjoyed even in the big frosts.

He had a word about the people who commonly made use of pots filled with charcoal to set in the greenhouse, which was highly dangerous, he warned, to the people who attended to them as he had known them be almost suffocated. The method was also injurious to the plants.

Miller's drawing of his house gives a structure some 200 feet long with the centre portion bearing a room above it and two stove-houses at both ends. The glass is shown in small panes.

Being the methodical logical Scot he was, Miller goes into some pretty detail about his houses. For instance if it could be afforded, he says, you could wainscot your walls with wood, but he pre-

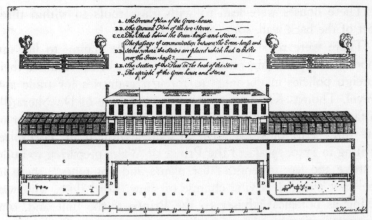

Miller's most substantial greenhouse with gardener's house above.
Note the "many bends" in the flued wall and the glazed tops of the
stoves abutting.
(Philip Miller's *Dictionary*, 1759 ed.)

fers they be plastered even behind the wainscoting, but, and here
again the practical gardener speaks, the woodwork should be
painted white, not black as he had seen them in houses where in
winter the leaves of the plants had fallen off.

Miller's ceilings were not half-and-half affairs either and were
to be lined with a proper thickness of reeds, heath or furze, laid
between the ceiling and the tiles, a foot thick. This was covered
with laths and given a coat of lime and hair to keep out the vermin. Many people had complained, remarks Philip Miller, that
the frost had entered their greenhouses during winter through
the front glass, whereas it came via the roof which had not been
furzed, lathed and plastered and tiled.

Philip Miller goes on to describe the two kinds of stove-houses,
the dry stove for "aloes, cereuse, euphorbiums, tithymals and
other succulent plants which are impatient of moisture."

Then there was the bark stove, with a pit carrying tanners'
bark some six or seven feet wide and three feet deep for the
reception of "the most tender exotic trees and herbaceous plants,
which before the use of bark it was thought impossible to be
kept in England."

These houses were lean-to with glass roofs to within three feet of the back wall, some sixteen feet high.

There were royal patrons too who were happy to lay out money and spend time on the new exotics, a direct result of their foreign policies in opening up foreign countries for trade and travel. Thomas Knowlton, who was in charge of Dr. Sherard's botanical garden at Eltham in Kent, and later head gardener to the third Earl of Burlington at Lanesborough in Yorkshire, writing in 1750, speaks of the Prince of Wales preparing to build a stove 300 feet in length "for plants and not pines"; there in the "not pines" is a vital change of view. Lord Bute, who was acting as botanical adviser to the Prince at the time and, one suspects, acting as sponsor to the royal enthusiasm for horticulture, arranged for new plants and seeds to be collected for the Prince from Asia, Africa, America and Europe, and "with the young Prince interested," as Knowlton remarked, "what could not be expected?"

In the North, Knowlton tells of his having just finished another big installation for William Constable of Burton, near Hull, Yorkshire, of two stoves, with a little greenhouse in the middle, 170 feet long, the longest and finest he had seen, but he does mention that a Mr. Salvill, at Crosedeal, near Durham, may equal it, although the one at Burton had more fire walls for vines.

Knowlton emphasised the early trend to use greenhouses for the cultivation of foreign fruits and the preservation during winter of exotic plants, when he says that pines were as common as any fruit and wondered what would be the next popular fruit or plant novelty of the gardening rich. Would it be, he says, the mangosteen (gamboge tree), with an orange-like fruit of chestnut brown, recently introduced from the Molucca Islands, or was it going to be jacks, the bread-fruit tree from India and Malaya, another recent introduction?

Shortly after the middle of the century Speechly, gardener to the Duke of Portland at Welbeck Abbey, built his huge pineries, some 250 feet long, heated by flues and having large brick tanks for plunging the fruit in tan or oak leaves. He advocated stoves for gentlemen as providing "indigenous vegetables and a variety of fruits and flowers, the natives of a warmer climate."

The Yorkshire-born Quaker, Dr. John Fothergill, one of the foremost botanists of his day, was fortunate in making a tremendous amount of money from a most fashionable and busy London practice. This was seemingly by treating his patients on a purely scientific basis rather than with quackery and remedies which should have gone out with the witch's cauldron.

As soon as he felt able, actually in 1772, Dr. Fothergill tore himself away from his lucrative London consulting-rooms and indulged himself in his favourite hobby, botany, allied, quite appropriately, with his professional interest in medicinal herbs and plants, at Upton in Essex where he set about the task of building up one of the finest botanical collections in the country.

A contemporary describing this really pioneer greenhouse man spoke of his fifteen gardeners and said of the Upton gardens "on the banks of a winding canal exotic shrubs flourished in the middle of winter, evergreens were clothed in full foliage without exposure to the open air; a glass door from the house gave entrance to a suite of hot houses and greenhouses, nearly 260 feet in extent, containing upwards of 3,400 species of exotics. In the open garden nearly 3,000 species of flowers and shrubs vied with natives of Asia and Africa."

The botanical doctor kept three or four of the best artists in almost full-time work to draw the new plants as they came into flower and fruit, so that should any disaster happen later he had a permanent record of these rarities. Indeed when he died twelve hundred natural history drawings from his collection were sold to the Empress of Russia.

It was further said of him that "whatever plants obtained a place in the materia medica or promised to be of service in physic, or manufacture or was in any way remarkable for rarity, beauty or physiological habit, was sought out and bought without regard to expense, and no pains were spared in its cultivation."

Correspondents in all parts of the world were continually sending him new plants and seeds, one of these, Peter Collinson, of Mill Hill, London, had the American amateur botanist and plant hunter, Bartram, searching the virgin forests of the New World for the enthusiastic doctor.

How could such a rich man or his friends and acquaintances escape their duty to horticulture and the great goddess Flora when there were impassioned appeals like that of Richard Steele in his *Essay Upon Gardening*, printed at York in 1793?

"My present intention," he wrote, "is an attempt to aid in the management of that most elegantly refined and fascinating department of the garden where the prodigious varieties of rare plants that have been introduced into this Kingdom from the hot regions of the terraqueous globe are deposited.

"With great deference, gentlemen, I will appeal to you if it not be rather incumbent upon you who reside at ease in this our favoured island and enjoy the pleasant sunshine of individual fortune to cherish and preserve those great curiosities of the vegetable kingdom which the bold and adventurous have, with extreme hazard and difficulty, collected.

"To obtain these rarities men of the greatest accomplishments navigated unknown seas, have traversed drear isles and deserts, searched the forests of both the Indies and explored the burning countries of the torrid zone."

What man of taste, of feeling, what patriot, I say, could resist such a claim? Books were pouring off the presses now telling how best a gentleman of taste and feeling could follow this advice, while neighbours and friends were scrambling to erect the best stoves and greenhouses according to the most modern designs and ideas so many of which were available.

There was George Tod, with his *Plans and Elevations and Sections of Hot Houses, Greenhouses, an Aquarium, Conservatories, etc.* printed at York in 1807, writing of greenhouses he had designed in all parts of the country at the turn of the century.

He had learned the lesson of glass roofs for all his houses and in Tod's book we have descriptions and drawings of many fine houses for the higher classes of society, who, said Tod, had lately taken up botany, an elegant and interesting study, as a favoured pursuit.

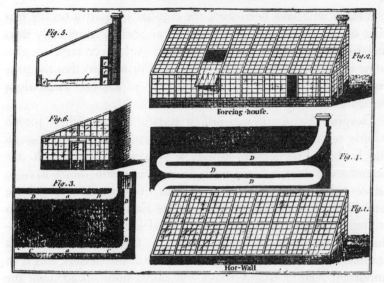

Another plate from Abercrombie of a typical forcing-house with the lowest of front walls and four flues, three in the back wall and one in the front. The other illustration is of a "hot wall" forcing-house with flues in the back wall itself.

(*The Hothouse Gardener*, John Abercrombie, 1759.)

Heating in these houses was still by hot-air flue. There was a hot-house for the Duke of Bedford at Woburn with a pit for forcing roses, and a hot-house for His Majesty at Frogmore, Windsor Great Park, and a massive job for a Mr. John Anthony Bucher, of Wandsworth Hill, Surrey, consisting of peach-house, a vinery, a pinery and standard cherry-house 240 feet long and 10 feet wide, with seven stoke-holes. Can one imagine the expense of such a house today?

Lord Heathfield of Nutwell Court, Devon, had built two peacheries and a greenhouse 129 feet long on the lean-to prin-ciple; while a Palladian-like structure with a solid domed top, built for Lady Clarke, of Windsor Forest, was not to the liking of the busy Mr. Tod for, he said, it robbed the roof of glass, a constructional point he did not favour and said so, but Lady Clarke won the day for her dome which was in keeping with her mansion. The house was 130 feet long and 14 feet wide.

Yet, as will have been seen, the emphasis was still on the forcing of fruit and even the Horticultural Society in its early days had to be persuaded by Salisbury, an early secretary, to give some attention to ornamental plants and flowers so that the proceedings should not be monopolised by papers and discussions on fruit and vegetable culture only.

Despite this there was still a terrific pother when Joseph Sabine reported to the Society on the fruiting of the mango in the Earl of Powis's hot-houses at Walcot Hall. He told the assembled Fellows: "It is with great satisfaction that I have to address the Society upon the subject of mangoes which have lately been communicated to us by the Earl of Powis. And thus having succeeded in showing that this delicate fruit may be produced under artificial management in Great Britain in such abundance to form a not unfrequent part of the dessert, this is so important a circumstance that I feel called upon to acquaint the public as speedily as possible with the results of this very signal triumph of skill and perseverance over difficulties which have hitherto been considered insurmountable."

Nor when Walter Nicol in *The Garden Kalendar of Horticulture*, of 1812, writes an introduction is there much doubt in which way his thoughts were running. He pointed out that the reader could not help but observe the rapid extension of the forcing garden with its hot-houses, flued-pits, hot-beds, and fire walls for the forcing of apricots, cherries, figs, grapes, nectarines, peaches, pineapples, strawberries, asparagus, cucumbers and melons.

The great and the wealthy, he concluded, found a source of real pleasure, gratification and amusement by the production of fine, excellent fruit to a considerable degree of perfection and of many mature fruits and rare esculents at an early and untimely season.

5

TECHNIQUES—GENERAL

FROM Europe in 1714 came descriptions and engravings of what were obviously the prototypes of the modern greenhouse. In *Neue Garten-Lust* in two delightful engravings are shown early houses with tiled and boarded roofs, wooden walls and casement windows hinged to swing outwards their full length, a great advance on sliding sashes and a principle the builders and designers of greenhouses in this country did not adopt for many more years to come.

A small, low arched doorway, under the casements in the front wall to lead into the house, was apparently designed in miniature as it were, to lose as little heat as possible when it was opened. Plants are shown in the ground in rows and in beds and three substantial tiled Dutch stoves are shown against the back wall.

The second drawing shows a half-domed roof, apparently built of boards, but it could be thatch, and with casement windows reaching from eaves to floor, a distinct improvement for light if not for heat. In 1737 a Dutch publication shows houses, the improvements in which, when compared with the 1714 vintage, are quite astounding.

Here, and it is certain that the discerning visiting noblemen gardeners and their professional assistants would take due note, are greenhouses, which but for lack of top light, and not too much of that fault, would not altogether have disgraced the gardens of a century later. The houses illustrated were obviously fruit-forcing houses. They were lean-to types with steeply raking glass fronts almost floor to ceiling, indeed, almost a large heated frame on end as it were, and very like the earliest wooden green-houses at the Oxford Physic Garden.

Through the back wall ran a flue fired from both ends. The glass frames are in casement form and are hinged at the top so as

E
65

to lift up. There would seem to be no door so that entrance must have been made through the frames. The small area of wooden roof sloping up to meet the back wall is in sections the same size as the frames below them and each one is hinged to lift up for ventilation. A second engraving shows a house with a similar sloping front, not so steeply pitched but with more roof, and one which is not only tiled but underdrawn to form a ceiling. In this house the fire flues run underneath the floor, returning to the same end as the furnace. The windows are casement type with small squares of glass and each frame appears to lift out like a Dutch light. A most tidy arrangement of pulleys and rope allows hessian blinds to fall and cover the windows both for shade and heat conservation at night.

At Leyden the early houses there were followed by a conservatory, on the north side of the gardens, which was 40 feet long, 10 feet deep and 14 feet high. The building was finished in 1744 and with its heavy walls and ceiling insulated by buckwheat chaff made an ideal hibernating place for plants as it did not need to be heated. Showing something of the efficient construction of these early plant-houses, the present Curator at Leyden reports that in the severe winter of 1928–9 when an outside temperature of 4° F. was recorded the inside temperature still showed 30°.

Boerhaave, the curator during part of the seventeenth century until his death in 1730, used lean-to houses, glazed in the front and raking back steeply so that there was but little space for a roof. The Dutch free-standing stove was most probably used for heating.

This was progress indeed, and although such ideas did not percolate all that quickly from the Continent to England, come they did. It was not, however, until the middle of the eighteenth century was reached and passed that progress was sufficiently made to be recorded. For it was then that many of the first rank of the aristocracy and a few rich professional men attracted by the tremendous fillip given to botanical studies by foreign travel and discovery and very often prompted by the beautiful exotics they had seen when young and impressionable on the Grand

Tour wished themselves to have these foreigners in their own gardens.

If native wit and common sense did not dictate winter protection for the tender greens they wished to grow, then experience soon pointed the moral and, with Dutch examples in print to show the way to a reasonably constructed hot-house or greenhouse, and nearly a century's experience in the conservation of oranges and curious greens in orangeries and stoves, hot-houses and greenhouses began to be erected, houses which would at least do part of the job for which they were intended.

These houses were still far from the light, airy, glass-roofed structures we know today, but in them their botanising owners were able, by the most careful and skilful management, to keep the newcomers, the visitors from foreign climes, if not in a thriving state of robust health during the dark, cold days of winter, at least alive.

That was a tremendous advance in greenhouse practice, and slowly, very slowly as the oranges and lemons and such "tender greens" were losing their favoured place in the winter sunshine and warmth of conservatories up and down the country much more comprehensive collections were taking their place. That was always dependent upon whether you had the wherewithal, the money and opportunity to travel or a wide enough circle of friends or contacts over the seas to provide and arrange for the transportation of the new plants and seeds available.

The portable heating apparatus used at the Oxford Physic Garden, circa 1621–50.

Even more necessary was the unbounded enthusiasm to fight the uphill battle against what at times must have seemed an almost impossible task, that of preserving in a state of growth the flowering and fruiting plants from climes which were, in many cases, the complete antithesis of the British summers and, certainly, the British winters.

That there was ingenuity and will to fight the natural hazards was well shown by the Oxford Botanical Garden's heating apparatus as used in the early conservatory type greenhouses which, as has been seen, consisted of a four-wheeled oblong metal wagon filled with burning charcoal which, when the elements warranted, the gardener drew by a long handle backwards and forwards along the greenhouse path.

A pertinent voice and one which did much to start gardeners thinking that all was not well with their heavy stone, small-windowed, dark, ornamental orangeries and to turn their thoughts to more enlightened practices was Richard Bradley's.

It is particularly stimulating to read the opinions of this very practical Fellow of the Royal Society and Professor of Botany at Cambridge on early attempts at greenhouse and hot-house culture. In his *New Improvements of Planting and Gardening Both Philosophical and Practical*, which has a grand eighteenth-century ring about it, and was published in 1718, he brings his most knowledgeable point of view to bear on the problem from the standpoint of a gardener with dirt on his hands, albeit he was a Don.

"As there is nothing more difficult in the management of exotic plants," he wrote, "than the right understanding of conservatories where they are to stand in the winter, so I think it is necessary to prescribe such rules for the building and ordering of greenhouses as I have found to forward the welfare of plants and as have yet been but little regarded or understood. The greenhouses as they are commonly built serve more for ornament than use. Their situation to receive the south sun is the only thing that seems to be regarded towards the health of the plants they are to shelter.

"It is rare to find one among them which will keep a plant well in the winter either by reason of their situation in moist places, their want of glass enough in the front, or the disproportion of the room within them.

"And sometimes where it happens that a greenhouse has been well considered in these points all is confounded by the flues under it which convey the heat from the stoves. Besides what is commonly called a greenhouse, it has been customary to provide glass-cases of several kinds and stoves for the preservation of plants brought from different countries, but I now find they are so many unnecessary expenses; a good greenhouse, well contrived will do all that is required for the welfare of any plant in winter; it may be so ordered as to shelter at one time orange trees, plants from the Cape of Good Hope, Virginia, Carolina, and indeed such as grow within ten degrees of the line.

"I must confess when I was first acquainted with aloes, Indian figs and such like plants that they could never have heat enough and I destroyed many by that too common notion; I could hardly venture them out of the hot beds in the most extreme heat of summer, and in the winter half roasted them with subterranean fires which commonly raised damps, cracked the passages of the flues where it ran and oft filled the house with smoke, which is a great enemy to plants, especially the smoke of seacole."

There were inconveniences, dangers and even worse from the use of charcoal, said the learned professor. "Several men have been choked by them," he tells us, "and sparks from them have set fire to the house, but that depends on the care of the gardener."

But while all this technique was sound, Richard Bradley had not yet learned the vital value of ambient air and full light. A further description of a hot-house by him shows that he was writing about a lean-to with a "good thick wall" on the north and east sides with glass at the front to the south, excepting for a wall a foot high. The glass in this house was to be removable to let air in three weeks before the plants were to be put out during the summer, and three weeks after they had been put in for the winter. Even with his little charcoal fire he stipulated wooden shutters one inch thick to be shut every night.

At the Oxford Botanical Garden a progressive outlook had been maintained and had prevailed from the days of their first solidly built conservatory of 1632. Their additional plant-houses were built on the most enlightened lines and were, if their mode had been universally copied, correct examples for contemporary would-be plant-house builders.

At quite an early date, round about 1726, the first conservatory which had been turned into a herbarium was replaced by the eastern architectural conservatory with a gift of £500 by Dr. William Sherard. Wood in his diary, under the date 21 May, 1695, mentions he had seen a new herb-house, which in 1715 had had to be enlarged to receive a collection of curious exotics presented by Bishop Robinson. Between 1734 and 1736 two architectural conservatories and the first wooden greenhouses were built and about the same time a stove-house 30 feet by 14 feet by 12 feet high with glass only on the one side. The first wooden houses, pictured in the *Oxford Almanack* of 1766, had long narrow windows arched at the tops, a high, steeply-raking front, with the merest of opaque roofing more in the nature of a coping. The ends were solid; but here was a tremendous improvement, a house which, because of its all-glazed front reaching from the very short front wall to within a foot or so of the back wall, was almost all glass. According to early accounts they were 30 feet long, 10 feet wide at the floor level and 14 feet high. They were placed one at each side of the Danby gate and served until 1834.

Now, there were definite signs of progress abroad in the nation's gardens; lessons were being learned in the difficult and disappointing school of experience. Continental practices were not only being studied but copied and carried out.

Stephen Switzer, that apt and clever pupil of the great George London, the royal gardener, was as he writes in 1724 one of the first men, along with Bradley, to advocate more glass and less solid ceilings in this country. Then he published a plan for a forcing-house with this decided advantage, suggested, he said, by the grapery at the Duke of Rutland's home at Belvoir Castle. It can be accepted as a fact that it was not until this time, the first twenty or thirty years of the eighteenth century, that regular structures, only as yet of the lean-to type, roofed with glass and artificially heated, were seen in this country, but there were very few.

The Duke's graperies were noted by the Reverend Mr. Lawrence in 1718 as having sloping walls heated with internal flues from Lady Day to Michaelmas, but not until 1724 does Switzer record that the walls were covered with glass, and this, although it was little more than a glass frame, was among the first records of a heated, glass-roofed greenhouse in this country.

The sloping flued wall was not a great success, and it was on Switzer's suggestion that some of the flues were stopped up, some put under the floor, and the wall glazed to the top. Switzer was not sure about the sloping wall idea because of the dampness caused by the earth backing, although his illustration of a fruit-forcing house in his book *The Practical Fruit Gardener*, in 1725, shows a greenhouse with a sloping back wall.

Switzer gives a plan of this house which he suggests is an ideal one, and an elevation drawing shows a sharply raking back wall held up by rubble in an attempt to keep it dry, with a completely glazed front. The plan drawing is "of a house 40 ft. long, about 5 ft. wide from out to out and 3½ ft. in the clear of the inside, all of which is employed as a border to entertain the roots of the trees, except five or eight inches towards the front which I have allotted for a little alley to set some pots of flowers or other things that are perhaps more necessary in."

Small stoke-holes, three in number, are let into the wall, so that "a hat full of coals at each place" would be sufficient to heat the whole length and he advocated a cast-iron pipe under the border with a row of square tiles over it. This pipe, said Switzer, would heat quicker with less coal and keep hot longer than a brick flue. "What is of no less moment it will diffuse itself with more intenseness and to more purpose."

He advocated 45 degrees for the slope of the roof, which, from an engraving, was apparently slated. Owing to the slope of the back wall and the front being in parallel, however, the slated roof was only a matter of three feet six between the back wall and the front glass. This was a big step forward. In a further

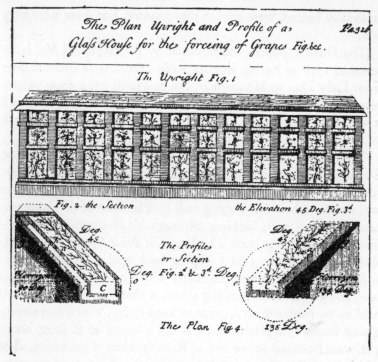

One of Switzer's early forcing-houses with more glass than any other house of its time. The two bottom rows of glazed openings provided for the windows to slide. It will be seen that the back wall was sloping, as well as the front, leaving a minimum of opaque roofing.

(*The Practical Fruit Gardener*, Stephen Switzer, 1724.)

elevation drawing which accompanies the text there are shown three rows of window openings, with the top, and smaller row, of boards.

Bradley also described some newly invented stoves and conservatories he had seen. One of these newfangled ideas was a niched wall facing south with two ovens on the back wall, the heat from these being led by flues in the wall right and left round the sides and along a low front wall to a chimney in the centre. On top of this very low wall and flue combined, panels of glass sloped gently towards the top of the back wall. This might well have been a further description of the Belvoir Castle vineries for vines were planted in the niches. The next stove or conservatory Professor Bradley saw was that at Hoxton, belonging to Mr. Fairchild. It was some 12 feet from the front glass to the back wall and about 40 feet in length. The front wall was only two feet high and from it there rose the wooden frame to which the glass frames were hinged. "This lies sloping to the back in such a manner as to drop about a foot from an upright. The roof or ceiling was higher by a foot at the front and sloped to the back wall."

An oven a foot high and three feet square opened into three flues running parallel to the front wall the full length of the house. At the far end the three flues merged, rose a foot or so and returned at the back of the house under steps topped with tiles. Over the front flue, covered with tile also, was a foot of sand put to conserve and regulate the heat. Bradley said by taking away or adding to the sand blanket heat could be regulated to the pots placed in it.

Over the oven three earthenware pipes were placed to allow cold air from outside to be warmed up before passing into the house to allay damps.

In some other stoves, said the eminently practical don, the oven was covered with plates of cast iron so that the space of the oven was about 10 feet in length and 5 in breadth. The flue from this ran parallel with the back of the house and then along the front just within the glass, ending in a chimney that passed through the end wall in which the oven was set. The flues were 14 inches wide, 18 inches deep and covered with iron plates about two feet in length. Sand to some three inches thick was laid on top and upon that was placed a pavement of tiles. That type of stove would do for a house 50 feet by 10 feet by 10 feet, and he advocated that Holland peat be used for fuel. He supposed by such stoves might be ripened (notice how all the emphasis is on fruit) ananas, cucumbers, melons, bananas, guavas and mangoes. He went on to say that at Badminton the late Duchess of Beaufort had ripened guavas by means of such a stove. He waxed even more enthusiastic after a talk with Mr. Telende, gardener to Sir Matthew Decker at Richmond, and seeing the pineapples raised there in glazed hot-beds and stoves. This particular stove used the iron-plated flues and had a thatched roof and shuttered windows. He burned nothing but turves on his fire, said the observant professor. Mr. Telende had a thermometer and of it the enthusiastic professor said: "By the same thermometer we find that the heat of the Stoves and hot beds may be raised much higher, even to be equal to the heat under the line or equator; so that there is not any plant upon earth which may not be made to grow in Britain by such help. The spice trees or plants of the hottest climes, whether the nutmeg trees, the cinammon trees, the several sort of clove trees, the pepper or the ginger will certainly prosper in such Stoves, if we give them their exact share of heat."

"A Country Gentleman" who wrote *Tracts on Practical Agriculture and Gardening* in 1769, had much to say, both cogent and interesting, about the greenhouse and conservatory. Of particular interest today is his estimate of the yearly cost of maintaining a stove 40 feet long in those days. If the value of the pound in 1769 is gauged by the fact that a man's labour could be got for 2*s.* per day, then one can the better see the cost the eighteenth-century gentleman incurred in following his botanising hobby.

First our country gentleman gave an indication of the popularity of the art, opening his discourse with the sentence: "As a garden is thought to be complete with a Stove for the raising of pineapples and as the building and looking after it has been magnified in the imagination of most people beyond what it really is, I set down every charge of my own for upwards of three years."

This was in 1764 when he thought his estimated cost should come within the pocket of "almost everyone settled in the country." The stove was 40 feet long, 12 feet wide with upright glass in front, and a solid roof. This had cost £80, as he was near London and the material used was new. If the stove were built against an old wall then £15 could be knocked off the cost. The fire was at the west end and the east end was glazed "for the

Thomas Fairchild's greenhouses, latticed-windowed with long sloping roofs, almost enlarged cold frames.

(*The City Gardener*, Thomas Fairchild, 1722.)

rising sun to shine through." The front was to have a south aspect.

The plants for the house—pineapples—would cost from 12s. to 15s. apiece if near fruiting time and from 1s. to 10s. each if they had to be grown on, but he averaged that about £30 would stock his stove with some 150 plants. Besides the pines there could be found room for 400 pots of strawberries and 80 pots of cucumbers or French beans.

The yearly cost of the stove is set out below:

	£	s.	d.
800 bushel of tan to fill pit at 1½d. a bushel	5	0	0
16 loads carriage at 12d.		16	0
3½ chaldron of coal at 36s. . . .	6	6	0
200 bushel of tan to keep bed level . .	1	9	0
Filling pit and planting the pots, two men two days each at 2s. per day . .		8	0
Stirring the tan up and putting fresh in, four times more at 8s. each . . .	1	12	0
For attending the stove 52 weeks at 8d. per week	3	18	0
Repairing the windows, painting and white-washing the stove yearly about .	1	11	0
	£21	0	0

He went on to say that that spring—1769—he had seen several stoves and fruit-houses in Holland and hoped he might have found some improvement in them different from ours. But they were not nearly so commodious nor so well contrived as the modern English oven, he commented.

If the grower of pineapples was commercially minded, said the country gentleman, who must have had a most mercenary strain in his make-up for a gentleman, the pines could be sold, and if the grower had them ready against the 4th of June, the King's birth-day, then he could have a price well above 8s. a pound "On account of the many entertainments which the Ambassadors and noblemen give on that occasion." To keep the cost of fuel down cow dung and coal could be mixed into fire-balls during the summer and dried.

Frosty weather technique was difficult and if the fire did go down it could mean a bushel of coal and one or two hours' work

Oxford Almauack 1766

A view of the Oxford Botanic Garden with the Danby Gate flanked by the two first wooden greenhouses built in the garden in 1734 at the same time as the two adjoining architectural conservatories in solid masonry. The wooden houses, 30 feet long by 14 feet high, were lean-to with steeply sloping sash-lights under round-headed top lights. They served until 1834.

(*Oxford Almanack*, 1766.)

for the gardener to get the heat up again. One method designed to alleviate this struggle in the early hours was to have the gardener fork up the tan-beds or to light six large candles.

As light was found to be so necessary, he continued, in his hints and suggestions, and Mr. Miller had proved this to be so, he advised that the unglazed end and the back wall should be covered with white Dutch tiles or else well plastered and painted white with the best gloss to reflect light and the heat of the sun.

"If the light at night was thought to be of use then a globular glass lamp with a tin funnel for carrying off the smoke could be hung from the roof of the stove for a cost of about 2*d*. per night for oil."

To save watering all the 400 pots the writer advocated a lead trough, into which the pots were put, connected with a water cistern, at which, with a turn of the cock, the trough and the pots were watered in a few minutes.

In *Gardener's New Kalender* of 1758 by Sir John Hill, the author spoke of stove-house practice in Holland where it was usual to keep "a little fire going in one corner of the greenhouse with a few mouldering turves to keep heat without violence for ever so long."

His advice and strictures on greenhouses and stoves generally were excellent for their day and sound horticultural practice.

"It is common to load a greenhouse with upper rooms, but that is wrong," Sir John advised. "The back wall may serve for the erecting of sheds for the coals but nothing more should be done. No superstructure should be allowed because it implies a solidity below which is out of the character of the building. . . . A greenhouse properly built," he went on, "is in character and is an ornament."

He was very definite about measurements and laid down 15 feet as the proper width and if, he warned, the care of plants was the sole consideration of the gardener then that measure had not to be exceeded whatever the length.

"The back is to be a straight, upright wall; and the front, in a manner, all glass. The height should be in front one-eighth more than the breadth, consequently the windows should be sixteen feet in length, and they should reach from the top to within ten inches of the ground, a wall of that height being raised to receive them.

"Here then is the form of the greenhouse. The back and ends are to be of brick, and there must be a low brick wall in front;

from this are to be raised sashes to the ceilings; these should be five-feet-and-a-half in breadth, and piers of brick must be carried up between them for the support of the roof."

Then followed a very detailed account of the interior fitting. "Every part of the house," said Sir John, "must be contrived for warmth and for defence against frosts. The floor should be raised 15 inches above the ground, that no damps may come that way and the next care must be of the sashes. They should be hung in firm grooves of seasoned wood-work; and their frames must be solid and well-jointed. They must be wrought so as to move easily, and to remain very firm in their places; no wind should be able to shake them; and they must be secured by sand bags wherever the air could get in."

The door, he advised, should be narrow, in the back or west wall and should be double so as not to let in the wind when it was opened. The floor was to be laid with paving tiles and the roof should be slated.

"As there will be room under the floor it will be advisable to carry a flue right through it with two returns. The fire place should be in one of those sheds we have directed should be behind, and that nearest the West end will be best. From this it should be carried to the front and along to the East end, and thence to the back again; here funnels must be raised to carry off the smoke."

Shutters were to be made for all the front glass and these had to be made so tight-closing that when they were shut on the brightest of days no light could be seen through them.

Plants were to be placed in the house in tiers.

Sir John spoke of the stove-house quite separately and distinctly as an "article much less understood and consequently much worse practised than the others."

In describing what plants to put in the stove he defined a greenhouse as a structure to hold those plants "that require a shelter in winter but usually no artificial heat" and a stove as those "that required both shelter and heat."

"The stove," he described, "is a building glazed in front and exposed to the south in the manner of the greenhouse; but it is to be glazed at the top and it must have fire places and flues in the back wall." A distinct improvement this with glazed roof! The house could be divided by glass partitions for housing plants requiring differing degrees of heat. Plants were to be placed on stands in the centre or in tan-pits, these two items being the difference between dry and bark stoves. As the top had to be defended both against rain and sun at times, either shutters or canvas blinds had to be provided.

If a gardener had his choice of bark or dry stoves, Sir John was sure he would choose the former as the harshness of dry heat was not natural and plants did not progress half as well as in the humidity of a tan-pit.

The stove was of lean-to construction, but the sloping glass roof singled it out as a most efficient horticultural building. With a length of 48 feet and a width of 18, two fire stoves, one at either end, were advocated.

A most clear description of preparing tan-pits is also given. To every load of bark, after the tanners have finished with it, are to be added a bushel of elm sawdust. The whole is thrown into a heap and allowed to drain and for the mass to begin to ferment. When the tan had lain a week it was thrown into the pit a little at a time and spread with care, spreading every parcel with a three-pronged fork and flattening it with the back of the fork. It had not to be pressed too tight nor left too loose and fresh elm sawdust was to be sprinkled upon each parcel. If the tan had been lying a fortnight out of the tan-pit and had lain a week in a heap then it would be some 15 days before it was warmed through sufficiently for pots to be plunged in it. Then with normal conditions the tan would preserve its heat for six months.

With the thermometers "now reduced to such exactitude and certainty that nothing is left to the judgment of the gardener but he is to do everything by observation, the degrees are marked and all he is to observe is, that he by no means let it rise or fall much below the true place," the gardener was to be given a tolerance of six degrees within which he must "absolutely" keep his bark stove.

Here was admirable advice for the taking but it was many more years before there were many such good stoves as that described by Sir John Hill, for some sixty years later, in 1807, Alexander McDonald in his *Complete Dictionary* still wrote of conservatories with tiled roofs and talked of building a good house for the gardener above the greenhouse. His actual definition of greenhouse was: "A sort of building fronted and covered with glass, destined for the purpose of preserving various sorts of exotic plants through the winter. In some the aid of artificial heat is not here necessary except in very intense cold weather."

It was but a few years after this when a whole dictionary would hardly have carried the theories, the learned theses and inventions which made up the body scholastic of the greenhouse art.

6

GLASS—THE BASIC NEED

It is almost impossible today to imagine the pother, the bother, the weird and wonderful ideas prevalent in the early days when a gentleman decided to set up a greenhouse, stove or conservatory as the next step in exotic culture from the orangery. He would probably have read what the Dutch had to say on the subject and what such experts as Evelyn, Bradley, Parkinson, Miller and Plat advocated and had learned from their own experience and that of their friends.

Besetting him on every hand would be innumerable books gushing over with professional and friendly amateur advice on all and everything about the greenhouse art. He would read of the "newest and only true improved method" of heating, multiplied to an almost insufferable extent with high-sounding claims for the greatly improved oven, the newly discovered stove, the patented hot-air house or the only recently discovered and invented improved system of heating. He could choose or flounder, depending upon his strength of will and determination, in a wrought- or cast-iron welter of tubular, conical, saddle, elliptical, calorific, portable, domed, gas, steam, water, upright, flat, compensating or terminal boilers.

He could decide, only after reading the most learned and abstruse discourses taking him into the rarified air of higher physics, which tinted or coloured glass he would choose, because if he believed all he was told or read he would certainly eschew just the plain ordinary stuff.

The question of whether to use metal or wood construction— still something of a problem today—was a most difficult problem in the nineteenth century, particularly with the magnificent example of the Crystal Palace as a highly successful lesson in metal and glass.

There was much learned discussion and donnish argument to consider before you could be altogether happy on the aspect of the intended house; and angle of roof fall was not a thing to be decided out of hand. As to the shape of your house, well—there was curvilinear, the acme of perfection (don't think about the cost), domical, ridge and furrow, polygonal, polyprosopic, lean-to, span, three-quarter span, semi-curved, octagonal or hexagonal coned, north light, cylindrical, square and oblong.

It was not easy to choose for it seemed at times that too many fertile and inventive minds were busy on glasshouse problems for the comfort and peace of mind of the keen amateur.

As Pulteney had said in 1790: "The increasing culture of exotics in England from the beginning of the present century and the greater difference of taste for the elegancies and luxuries of the stove and greenhouse, naturally tended to raise up a spirit of improvement and real science in the arts of culture. To preserve far-fetched rarities it became necessary to scrutinize the true principles of the art, which ultimately must depend on the know-ledge of the climate of each plant."

In reality there was much more, a great deal much more than that involved: height, breadth, length, the ventilation, and the heating of houses differed tremendously depending on what you wanted to grow—vines, plums, apricots, nectarines, cherries, heaths, carnations, malmaison carnations, succulents, figs, pines, melons, orchids, pot plants, ferns, palms, aquatics, orchard fruits, French beans, oranges, camellias, early strawberries, tropical plants or pelargoniums.

For this was the greenhouse climate of the nineteenth century. Gardening progress had been slow, but progress there had been and now houses like the above were a principal feature of the times, whereas during the major part of the previous century stoves and greenhouses were for the collector, for the botanist for growing the "guinea pigs" they were going to dissect for science; to enable them to put the vegetable world into its place; to codify, to classify, to catalogue, that was the prime urge. The love of flowers and plants for their beauty, colour, form, rarity and romance came later in the century, much later, when the insatiable eighteenth-century quest for knowledge and scientific enlightenment had abated.

The amateur botanist, the philosopher in the natural sciences of the seventeenth and eighteenth centuries who had found leisure employment, some a challenge, and some almost a religion, in their seeking out, preserving and close scrutinising of the almost indescribable wonders of nature which came under their gaze for the first time, must have found, as the plant-hunters roamed farther and wider and were ever more successful in their discoveries of exotics, that their hobby was demanding their full time, their unremitting attention to detail, and that their cataloguing became more intricate and complex as the floral wealth of the world piled up at the doors of their potting-sheds.

The medico-botanists who were, in the late seventeenth century and the first half of the eighteenth century, so much the sponsors and patrons of the efforts of the few plant collectors there were and the art of exotic cultivation and whose attentions had been directed to the uses of plants for medicinal and curative purposes—Bobart, Uvedale, Sharrock, Bradley, Richardson, Fothergill, the Sherard brothers, Collinson, Petre, to mention some—were not the arbiters of horticultural art or taste in the late eighteenth century or any part of the nineteenth.

Their place was taken by the newer aristocracy of taste who wanted beauty for its own sake, and later, in many cases, in the Victorian era, beauty as a background to living. And such practical honest-to-goodness writers at the close of the eighteenth century as Loudon and Nicol turned the scales completely with their insistence on sound methods of greenhouse and conservatory practice. Throughout the horticultural writings of the time, too, there was the fact, made patently plain, that any gentleman worthy of the name would obviously have a conservatory, attached to his "humble" mansion, and greenhouses, stoves, forcing-houses and frames in his garden.

After all, imagine a gentleman without glasshouse ranges! As soon hear of one without his carriage and pair!

Nicol, who was a working gardener and secretary to the Caledonian Horticultural Society, shows his sound Scottish sense in the following quotation on the construction of hot-houses, a quotation typical of the times:

"Various are the ideas entertained and the devices practised on the subject, and very far have some late schemes led the public who have held out a show of economy and persuaded many to alter well-constructed houses to mere gimcracks; with hot air flues and cold air flues; improved furnaces that set the house afire, by way of keeping up a regular heat, double roofing, inner roofings and much other nonsense too tedious to enumerate. But often, and indeed generally, soil and general management, proper heat, sufficient quantity of light and fresh air is the secret.

"The true position [of a hot-house] is to the sun at 12 and the most sensible line of front with a forcing house is straight not curved, both ends being glazed by which means every part in it may enjoy the full sun and light. Some argue for concave and others for convex fronts and insist that the plants enjoy more heat and that the sun shines more forcibly on the glass being more at right angles with it in his motion, than as a straight front.

"This is so far true, but the shadows within the house are broader and the different parts are longer shaded by one another, the more convex or concave the front is, so that a straight front, or one very little curved is to be preferred, for purposes where all the sun and light that can be obtained is desired." Clearly the job of constructing a glasshouse was being taken very seriously indeed!

Nicol, talking about greenhouses, in his own wording "a gimcrack" and a very different kettle of fish in his day to a hot-house, is brutally frank on the subject. "This compartment [the greenhouse] being an object of taste alone," he says, "is more subject to diversity in its cost (and that too with more propriety) than any other in the garden, and although the gardener should have the sole direction in building the different species of forcing houses, yet the fancy of the proprietor in respect of the green-house or conservatory may be more safely indulged in; since nothing is at stake here [note his utilitarian approach!] in comparison to what is in the pinery, grapery or peach house; the construction of the fireplace and flues is of importance here as well as in the hot house, and as the health of plants in winter depends much on the dryness or dampness of the floor I would recommend that the flues be run under the basement all round."

He went on to advocate that the floor of the greenhouse be paved.

"Air and light in winter," he said, "is of the utmost consequence to the health of plants, so that the free admission of them should be studied in the construction and at the same time that wherein a great deal of fire heat may not be required which tends to draw the plants up weak in winter and spring."

Where elegance was studied then freestone could be used in the construction of the house, but not, please not, in the Gothic style, although Nicol admitted masonry could have full sway in any of the other orders without much inconvenience to the plants.

Nicol wrote with evident disquiet about the eighteenth-century attempts at glasshouse building which he saw around him when he was writing his *Garden Kalendar of Horticulture* in 1812. Referring to these brick-pier or masonry fronted, flat- and opaque-roofed houses he says: "Such primitive greenhouses were general up and down the country. Such may be found in almost every county, many of which look more like tombs or places of worship, than compartments for the reception and cultivation of plants which ought always to be bright, airy and cheerful."

Nicol attempted to modernise some of these defects by stripping the roofs of slate or lead but "their great height" militated against their efficient use for they drew the plants in them up into thin, spindly objects.

From 1800 until 1860 arguments and theories on every conceivable aspect of covered garden building were rife, tempers waxed high, scientist confounded scientist, and one of the first items to come under the closest scrutiny was glass. Glass, the methods, cheapness or high cost of manufacture, its quality and its availability had always dictated, willy-nilly, the efficiency and the spread of the greenhouse art.

Throughout most of the seventeenth century and until 1845, when the tax was repealed, glass was subject to some form of direct or indirect taxation and its price kept at a high level, so that cost was, of course, a big factor in the popularity or otherwise of the cultivation of exotics in winter shelter. For the aristocracy throughout the eighteenth century and for a wider circle of greenhouse gardeners in the nineteenth up to the tax repeal, it was this high cost of glass which dictated to a large extent whether a gentleman could afford to have glass on any scale worth while as part of his garden establishment.

At the opening of the seventeenth century when a gardener's day rate was 1s. 6d., a tradesman's 2s. and an agricultural labourer's rate was round about 7s. 6d. a week, crown glass squares were 9d. each and as these squares could not have been above 9 by 20 inches, and were most probably less—perhaps an average of 6 by 8 inches—the cost of erecting a stove or greenhouse can be imagined. Towards the end of the century when an agricultural labourer's weekly wage had risen to 10s., crown foot squares had jumped to between 1s. 1d. and 1s. 5d. with thirds at 10d. Common glass bought in footage, but only in short lengths, to cut on the job, ran out at 7d. to 8d. a foot.

The greenhouses built from this glass, which was of a greenish tint, were built up of sash windows which pushed down or up over each other and were of small panes. At first the glass was fixed into lead casements, but later fixed into wood with either wooden beading or putty, which was first mentioned in 1737 in a book by Pierre La Court on domestic architecture.

Glass then was a determining factor in the evolution of greenhouse practice and until the repeal of the glass tax its high cost always militated against the universal popularity of the art. The repeal put a different complexion on greenhouse work altogether. Previous to 1845 a crate of good crown was £12, in 1865 it could be bought for £2 8s., while ordinary sheet prior to 1845 was 1s. 2d. a foot and fell to 2d.

Apart from cost there was also the question of manufacturing method which played its part in the evolution of the greenhouse. For several hundred years the greenhouse builder had no choice of glass but broad and crown, neither ideal for horticultural purposes and both types marred in the processes of manufacture. Broad glass, for instance, was made by dipping a cylinder into molten glass, cutting the glass-walled cylinder as it cooled and ironing it out flat. This method gave a glass which in total before trimming and cutting seldom exceeded four feet square, had unevenness of surface, that might well be ridged, crusted, crumpled and was never perfectly flat. There was also inequality in the thickness of individual panes which could vary from one-sixteenth of an inch to three-quarters. Inequality of texture made for air-bubbles, knots, streaks and varying shades in the glass too.

Crown glass which was spun into a circle at the end of the glassblower's pipe was, on the other hand, brilliant, of a greeny cast and well favoured by gardeners, but could not be cut into large panes as the average diameter of the glass spun was only from 50 to 60 inches, out of which had to be cut the bullion (the bottle glass bulge in the centre) and the rough selvedge. The panes cut had normally a slight curvature which caused some distortion in the light passing through them. It was obvious that all glass manufactured by these two methods would have to be used in many-paned window sashes with the consequent loss of light. When in 1833 new methods of manufacture resulted in good-quality sheet glass becoming available Loudon said that the introduction of this glass into hot-houses was "one of the greatest improvements made in their construction since the substitution of roofs of glass for those of opaque material."

Repeal of the tax gave an even greater fillip to both the manufacturer and the greenhouse builder. Improvements took place all round in size, quantity and prices of glass, and Paxton, who was then head gardener at Chatsworth, said of repeal that it "gave an impetus to horticulture which only a short time ago no efforts could have called into action."

The sheet glass now available was a little thicker than crown but could be obtained in lengths up to 6 feet, although the normal sizing was from 2 to 3 feet long and 1 foot wide with a weight of between 18 to 26 ounces.

And while until repeal there was little if any choice, except crown and broad, for horticultural purposes, in 1853 a contemporary writer speaks of British plate, patent plate, rough plate, patent rolled rough plate, crown, British sheet and Belgian sheet; crown, British sheet, and Hartley's patent rough plate, being the most popular and most used for horticultural purposes.

The price of crown, of which five qualities could be bought, was, from 8 × 6 inches up to 10 × 8 inches, 2*d*. a foot. Second crown, not exceeding 14 × 10 inches, was 7*d*. a foot, while British sheet up to 40 inches long varied with quality and weight from 3½*d*. to 1*s*. a foot. Rough plate of ¼-inch thickness up to 35 inches long was 1*s*. 1*d*. and patent rough plate much used for conservatory and greenhouse roofing cut to the size of 8 × 6 inches was 4*d*.; to cut, 30 inches wide and from 40 to 50 inches long, the cost was 6*d*. per foot.

In 1853 with cheaper glass, whereas hitherto, as a contemporary writer remarked, even the most opulent had been prevented by its price from enjoying many of the most delicious fruits of the tropics, now they would have the guava, custard apple, banana, bread-fruit, coffee, Jack-fruit, lee-chee, loquat, mango, mangosteen, and plantain in greater perfection than in their native countries, as with the pine.

The difference in light intensity to all greenhouses, as a result of the new and better glass, its greater transparency, and larger sizes, which did away with so many dust-harbouring laps and light-effacing glazing bars, was quite phenomenal, but it raised new and difficult problems. Sheet glass, it was found at first, caused quite considerable sun-scorching of plants.

The scientists had to get to work and the great Dr. Lindley applied his mind to the problem. First of all he warned against poor-quality glass and gave simple ways of testing, such as holding the glass some eight or ten inches away from a white sheet of paper and watching for luminous spots or stripes on the paper as the sun shone through the glass on to it.

For, as Dr. Lindley so carefully explained, this meant danger to plants. "If for instance," he said, "the roof forms such an angle that the sun can strike it perpendicularly near the middle of the day at the season when the leaves are young and tender, an effect may be produced by rays of light, however imperfectly concentrated by the irregular lenses, which effect would not be produced by the same glass placed at another angle of elevation as plants may be entirely out of the foci of the lenses." From the way sheet glass was made, he added, it must necessarily have numerous concavities and convexities some of which, or many as the case might be, had the power of concentrating the rays of light enough to burn the leaves of plants.

Just at the time the Palm House was being built at Kew in 1844–8 this question of glass physics was engrossing the attention of the many thousands of people concerned at that time in the building of bigger and better greenhouses, conservatories and hot-houses.

The new glass, its cheapness, the newer methods of building, iron construction, ridge and furrow roofing, and the great improvement in the technical arts generally, coupled with a greater distribution of wealth among a new middle class (determined to be as socially correct as the upper-classes with their horticultural amenities) had provided a tremendous stimulus to glasshouse building. Great ranges were going up in all parts of the country, and Paxton himself started the new hot-house at Chatsworth with the new glass, but while that great gardener was of strong enough character and of such intelligence as to work out his own problems, and apply his own answers, others would not move without due attention being paid, and weighty deliberation being given, to the many controversial ideas which swept in and out of the horticultural press and learned society transactions like so many swarms of bees on a high summer day.

There was that very weighty problem of aspect and roof angle, there was colour of glass to study, there was ventilation, now almost a new science for glasshouses, constructional methods and, of course, ways of heating in which it seemed every engineer and every plumber was working on his own particular invention of a newfangled furnace and circulatory system.

The colour of glass proved to be a tremendous object of scientific discussion and experiment when the new and bigger sheet glazing came in. Sun-scorch, as has been previously observed, was proving a new and irritating concomitant of the new glazing, and while the great Palm House at Kew was being built Mr. R. Hunt found it expedient to read to the British Association a paper based on what type of glass was most suitable for covering the biggest glasshouse in Europe abuilding at the time.

Mr. Hunt's theories so captured the imaginations of Her Majesty's Commissioners for Woods and Forests that they asked him if it would be possible to cut off the scorching effects of the sun's rays which they anticipated in the new Palm House by using a tinted glass which would not be objectionable in appearance. After an elaborate number of experiments he advised a glass coloured a pale yellowish green by means of oxide of copper. This glass was indeed used, in the Temperate House and the Fernery as well, but taken out later as proving more correct in theory than in day-to-day practice. And when colourless glass was used the effect was soon apparent in the improved vigour of the plants.

M'Intosh, writing at the time, said all these experiments did little but take his mind back to the days when good crown glass of greenish hue was in high repute for hot-house roofs.

Dr. Horner, of Hull, was another gentleman interested in experiments dealing with tinted glass for greenhouse purposes. He recommended violet-coloured glass "As not only affording partial shade but as transmitting a light which possesses a subtle action in exciting vegetation and proving in all respects an admirable auxiliary to heat and moisture necessarily employed in culture." His experiments must have been very different from most other people's experience, as violet-coloured glass, apart from any which might have been used in the Victorian patterned friezes to conservatories, was not taken up.

Mr. Hunt in his paper to the British Association had roved widely and certainly had given the subject of glass for greenhouses a much-needed airing. He found that by the use of red, blue and yellow media, the natural conditions of plants could be altered. He went on to give the differing results caused by the use of each different colouring. Light, he said, which had permeated a yellow media was found in almost all cases to have the effect of preventing the germination of seeds and, in a few cases affording the exception, the young plants had soon died. He thought that was caused by the action of the sun's rays and red light.

Light which had permeated a red media was not unfavourable to germination if the seeds were kept sufficiently moist to make up for the increased evaporation, but the plants under it assumed a sickly appearance and became partially etiolated so that the production of chlorophyl was prevented. Plants, however, it was found, remained longer in bloom under red light. Plants bent themselves as much from red light as they bent towards white light in an ordinary house.

Permeated blue light accelerated in a remarkable degree, Mr. Hunt told the British Association, the germination of seeds and the growth of young plants, but the rays were too stimulating and growth proceeded too rapidly without the necessary strength. If these quickly growing plants were caught in time and taken into yellow rays, or light which had permeated an emerald-green glass, carbon deposition was accelerated and woody fibre was formed in a perfect manner.

In orchid-houses and some conservatories a greenish tinted glass was used, but on the whole, while these learned discussions did good by drawing attention to the importance of glass generally and light in particular, ordinary window-type glass was generally used in glazing.

The correct angle of glass inclination to the sun for forcinghouses and greenhouses was a veritable bone of contention from about 1807 to 1810 between Thomas Andrew Knight, the president of the Horticultural Society, later the R.H.S., and the Reverend Thomas Wilkinson. The Reverend Mr. Wilkinson seemed to favour 45 degrees, but Mr. Knight was for 34 degrees for grape-ripening in July and 28 degrees for peach-ripening about midsummer, in other words angles for course!

At Leyden at the opening of the eighteenth century Boerhaave directed that his terms of reference to preserve plants during the winter required that the surface of his glass should be perpendicular to the sun's rays at the shortest day when most heat and light were required. His studies and data led people in this country, such as Miller at Chelsea, Dr. Richardson of Bradford, and the Sherards in Oxford, to decide on an angle of 45 degrees for the roofs of their stoves and conservatories.

Miller at Chelsea in his plant-stove required upright glass to meet the winter sun nearly at right angles and a 45 degree roof slope for the summer.

Formulae, which presupposed a reasonably alert mathematical facility, were put forward and a favoured one read: "Make the angle contained between the back wall of the house and the roof equal to the complement of latitude of the place, less or more the sun's declination for that day on which we wish his rays to fall perpendicularly. From the vernal to the autumnal equinox, the declination is to be added, and the contrary." An example put forward for Mr. Knight to ripen his grapes in the London area in July read:

Latitude of London	$51^\circ\ 29'$
Sun's declination on 21st July	$17^\circ\ 31'$
	$33^\circ\ 58'$ or
	34° nearly

Mr. Wilkinson made great play with his theory that "If the inclination were 45 degrees, the sun's rays would be perpendicular about April 6th and September 4th. And as the rays would vary very little from the perpendicular for several days before and after the 6th of April and September 4th, the loss of rays arising from the reflexion would be nearly a minimum."

By and large, however, while the scientists argued about their theories of reflection the gardeners built their houses with roofs of 45 degrees, an angle most likely, as even Mr. Wilkinson admitted, to shed rain efficiently. It did not matter even when M'Intosh, with his Scottish love of exactitude, complained bitterly that hot-house builders in general contented themselves by determining the length, breadth and height supposed to be most convenient to their own set of circumstances, not troubling themselves further or without going into the mathematical calculations necessary to arrange the slope of the roof to the latitude of the place, or the purpose for which that roof was intended. Hence he found the same angle of elevation in Cornwall and in Ross-shire, the difference in latitude and the sun's inclination being seldom worried about at all.

There was a further item too concerning the pitch of the roof with which the busy raisers of glasshouses had to contend. There was the loss of heat by radiation and conductivity all depending on the angle of roof slope. Indeed the scientists and the theorists had come into their own in the horticultural field now.

Mr. Perceval wrote learnedly that while radiation loss must be the same at all angles, the cooling effect of wind would be greater on a high-pitched roof than on a flatter one, though he added, almost as an afterthought, that he believed the difference in this respect, except as it increased the length of the rafter and therefore the cooling surface of the glass would generally be too small to be of any practical consequence. Mr. Perceval, however, was nothing if not thorough and he had worked out his formula that the cooling effect of wind on glass was as the square root of its velocity. No wonder the head gardeners of Victorian times were always pictured as highly serious men!

When you had decided on the angle of the glass and its colour there came the question of how it was to be glazed. And once again at the beginning of the nineteenth century the ideas on what today appears to be a pretty straightforward job was a most complex exercise in physics. The unevenness of the glass in the earlier buildings with its consequent air gaps between panes at the laps had meant that some form of stopping up the laps was necessary for which both lead and putty were used. For lead lapping the choice was between four different types, "s" bends, angle bends, Barrat's method, Saul's method and the ordinary method, one supposes, of just letting the glass lap and leaving it at that.

Then as to the shape of the glass—well, Loudon put forward seven differing shapes in 1825. There was fragment glazing in which all the old glass was used up, the panes being rough cut to

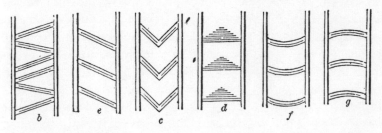

The various forms of glazing as put forward by Loudon: (a) the common rectangular (not shown), (b) the fragment, (c) the perforated shield, (d) the entire shield, (e) the rhomboidal, (f) the curvilinear, (g) the reversed curvilinear.

(*An Encyclopaedia of Gardening*, J. C. Loudon, 1825.)

all manner of odd shapes; the rhomboidal with the glass cut on a slope to form that geometric figure—with the advantage, it was said, of running water down to the lower angle of the glass on the sash bar; then the curvilinear with the glass cut to a convex lap with the tendency to conduct the run-off to the centre of the pane; the reversed curvilinear with the glass cut concave with the idea of throwing water equally on to both bars; the perforated shield glazing with the glass cut to a central point like a shield in reverse; entire shield like the latter but reversed; and, of course, the normal rectangular.

There was also on the market a "corroborated" glass. This was arched, seven inches broad and twenty inches long, with its own flange moulded instead of laps, this, its makers claimed, would not scorch plants; and also Russell's patent glass tiles, hung like a domestic tile and needing no sashes but only rafters. Several glasshouses using this method were built in Scotland.

7

HEATING—A MAZE OF IDEAS

THE efficient heating of greenhouses was a long, painful, and quite often a disastrous evolutionary process, exemplified by successions of scorched, "boiled," asphyxiated plants, and not a few asphyxiated gardeners too as the fumes rose thick and rancid from the old flue systems.

We have seen how the sun's heat trapped under sheets of mica in Roman times was the first recorded attempt to bring artificial aid to plant protection and growth and that the use of some glass erection was next recorded in France where it was written that flowers were grown in glass pavilions turned to the south.

Then at the first Botanic Gardens in Europe at Padua in 1545 there was built a greenhouse, probably of wood, rough masonry or brick, thatched with straw or tiled, into which plants were moved for winter protection. Rough mats of canvas or rush would go up over the opening in the masonry or wood to retain during the cold night hours some of the sun's heat which had been stored in the structure during the day and which would be slowly radiated during the night. It is probable that some form of heat would be used in the severest frosts, but there is no evidence to suppose this would be more than a pile of peat smouldering in a corner or in a brazier type of fire.

The forcing of fruit and vegetables was always of prime interest to the French and to them must go some of the credit for pioneering the way, although they fell behind in the real race of conservatory and greenhouse floriculture as it developed later.

It is recorded, for instance, that cherries were ripened at Poitou early in the sixteenth century by artificial heat, hot limestone and hot water being placed under the trees. Fruit from these trees was sent to Paris on May 1st.

Bernard mentions that arcades opened to the south were first

G

erected at St. Germain for Henry IV round about 1570 for accelerating the fruiting of peas.

Next in the record is the greenhouse in 1648 of Jungermann, the celebrated botanist in charge of the Altdorf University Gardens in Saxony where it is almost certain if heat was used it could have only been in a primitive way, for sixty years were to pass before Dutch horticulturists recorded orangeries and greenhouses with opaque roofs using either Dutch stoves or flued walls. The Dutch stove, still to be seen on the Continent, was a free-standing, rectangular metal structure on short legs, tiled above the firebox where the hollow of the stove and the flue-pipe itself, which passed straight out through the roof, formed a reservoir of hot air to radiate into the house.

Of course, these early gardeners were still engaged on a plan which ensured that they kept their plants in the open as long as ever they could without frost damage and housed them for as short a time as possible before they let them taste the spring again.

And that is why the so-called stoves, or the conservatories, had but a hole in the floor of the house in which to place a few burning embers. One wonders with such methods of heating how the gardeners of the period managed to follow the instructions of Samuel Trowell, Gent, as he called himself, of London, who in 1738 gave his ideas on the heat required for the various plants conserved.

In *A New Treatise of Husbandry, Gardening and other Matters Relating to Rural Affairs* he said: "It is likewise the part of the gardener to keep such heat in his house as will preserve his plants in the extreme cold; for many of the artists keep so great a heat in their houses that it draws their plants too much, which makes them weak and sickly and then any little check makes them ready to expire and very often die. Therefore 'tis conceived no house ought to be hotter than those months are wherein they (the plants) may safely stand out, which may easily be known by placing a barometer in the house (except some of those foreign plants that come from extreme hot climes as the banana, etc.) which require more violent heat than our climate admits of."

Seeing that Fahrenheit, the German, did not invent his mercury-filled thermometer until 1714 this would be much easier said than done. Not until there was available such a precise instrument was there any sort of certainty as to the degree of heat or cold of a house, and one has only to imagine where a modern greenhouse-owner would be without a thermometer to know what an exceedingly difficult job the old gardeners must have had to preserve anything but a native dandelion.

In Holland, alongside the flued walls and flooring and the free-standing stoves, it was also horticultural practice to have small fires of smouldering peat turf kept going in the corners of houses. This system was also used in this country, and even as late as the middle of the eighteenth century a writer of horticultural tracts advocated the heating of orangeries by peat burned in an earthenware pan or an iron kettle with a handle to move it about, a tin pipe to carry off the smoke and a piece of tin like a dish cover at the kettle end to catch the smoke.

At the Leyden Botanical Gardens, where the early plant shelters and conservatory were followed by greenhouses and stoves, Boerhaave experimented with heating from 1703 to 1730, probably using the Dutch stove or flue system, for that celebrated botanist was receiving exotic plants from the East Indies and members of the Geraniaceae and the Ficoideae from the Cape for the first time, all of which needed warmth and protection in the winter.

It would be obvious that any botanist, as the early greenhouse men all were, trained so thoroughly as they had to be in the Latin tongue, would be well acquainted in their reading with the hypocaust of the Roman villa and baths in which fire heat was led along tiled and bricked chambers and flues immediately under the floors of living-rooms and the sweating-rooms of baths. So much was the early flued systems a copy of the Roman method that in Rees' *Cyclopaedia* of 1819 the word *hypocaustum* is defined as "among the moderns the part or the place where the fire is kept that warms a stove or hot house."

And it was the ovens and flues built in the thicknesses of the walls or the "subterranean" ones of earthenware, iron or brick which were the main methods of heating in stove-houses of any pretentions at all, from John Evelyn's time—1660 onwards—to the full run of the succeeding century. Flues in the walls of buildings lasted for even longer in unglassed fruit walls until the nineteenth century, but for greenhouse work their day came in the eighteenth century. Peter Kalm, a Swede and pupil of Linnaeus, visiting Chelsea in 1748 noted "in the largest orangery in Chelsea Garden the smoke makes six bends in one of the long walls before it escapes."

In this comparatively simple and quite uncomplicated method of heating there was obviously vast scope for improvement and improvements were made. Horticultural writers of the time were particularly contemptuous and critical of the old methods so that some other method had to be tried. The subterranean fires, as so many of the old writers called them, received quite a broadside. There was indeed a reaction against heat at all and Bradley, from Oxford, thought that subterranean heat was much overrated and that a gardener could well do without it. That all depended, of course, on what the gardener had to grow and what he had to protect, and it may be that in those days, the first few years of the eighteenth century, not too many gardeners had a great need of heat for their few evergreens and citrus-trees.

For those who did need heat the first improvement was the use of pipes made of brick-earth, tapering towards the end, about 2 feet in length and 10 inches in diameter, fitting one into the other and placed on top of the ground; the next improvement came quickly after it was found that ground flues raised dampness and would not draw at all well. It was then found expedient to incline the pipe-line, raise it from the ground on chairs, to use pipes like present-day drain pipes made of fireclay and glazed, fitting by flange one into the other and making for much more secure jointing. Slight amendments on this system were the manufacture of square, narrow tile-like flues, egg-shaped and round-topped-ones, all advocated as the best yet and all said to have tremendous advantages over their immediate predecessors.

There were trials with pipes communicating directly with the

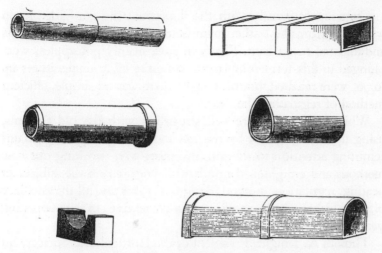

A number of earthenware flues, from the first used brick-earth tapered tubes to the various-shaped fireclay tubes, including a fireclay chair for raising the flues above the ground.
(*The Book of the Garden*, C. M'Intosh, 1853.)

cold air outside which led directly over the tops of the furnaces that fed the flues and were claimed to bring heated air into the houses and abolish damp vapours.

The heating systems used with the pipes was that of the fireplace in the outer wall which fed hot air and smoke into what was no more than an elongated chimney under the floor, or in pipes above the floor or into the walls themselves before debouching into the chimney proper which might be in the middle of the back wall, by the side of the fireplace or directly opposite the fire at the other end of the house.

The disadvantage of these methods of heating was the difficulty of keeping up a steady flow of heat and regulating it. There was too in the indoor stove the added danger of noxious fumes which could do and did irreparable damage to plants.

In both systems the equal dispersal of heat was never achieved and there was always the possibility of water freezing in a pan near the outside front wall, while plants were being scorched above the flue or near the back wall.

Contemporaneously with the flues went the tan-bark pits in which tan from the leather manufacturers was placed in deep pits and allowed to ferment. Plants in pots, mainly pineapples, were plunged in this fermenting mass and quite high temperatures up to 90° were reached, but once again there was no simple, efficient method of regulating the heat.

While gardeners were still struggling with the old systems, using up coal and peat by the ton and having to give their unremitting attention to the job, inventors were working out new methods and enlightened minds were engaged on the subject of heating, regulation, method, design of stoves and all the ancillary bits and pieces which arose in such an inventive era as the opening years of the nineteenth century.

Thomas A. Knight, President of the Horticultural Society, hit a very vital nail on the head in his researches and experiments when he criticised the too high temperatures raised in hot-houses as being due to a wrong principle based on men "being fully sensible of the comforts of a warm bed on a cold night and of fresh air in a hot day, the gardener generally treating his plants as he would wish to be treated himself and 'keep their feet warm'." This anthropomorphic attitude is still with us.

The first great improvement of the nineteenth century, an age chock-a-block with them in the glasshouse world, was the designing of detached, above-the-ground flues of brick on edge, 9 to 12 inches wide and 14 to 18 inches deep, or of specially moulded tiles, thick on the edges for strength and hollowed out in the middle for more heat, topped with paving-stone, cast iron or tiles hollow in the centre to carry water to give humidity.

These flues, designed to go round the house by the front wall returning by the back wall, at least had the advantage of radiating the heat into the house rather than into the thicknesses of walls or floors.

Mr. Loudon busied himself with the flue principle and suggested an improved one, divided into separate compartments, with a baffle at the end of each, the idea being for each compartment to expend as much heat as possible before the hot air and smoke passed on to the next. A hot-air pipe communicating with the outside air passed along the top. It was not successful.

Triangular cast-iron flues as well as various-shaped earthenware ones also were all tried out but still lost their heat too quickly and were too hot when they were hot. Other improvements were to the furnaces or ovens as they were still called. In this department many inventions and patents were introduced, still for use with flues, but basically they dealt with ways and means of introducing a regulated stream of air both through the fire and ash-pit doors to keep the fire in a healthy state without drawing cold air into the flue. Called a register or air-valve, although it was only a ventilating hole which could be opened or closed or partially so, at will, it was quite a step forward and shows, in terms of modern progress, what would appear to have been a particularly tardy development in fire and heat control.

A double furnace door and the registers mentioned were the first steps towards regulating the heat but scientists and engineers concerned with efficient heating soon developed many more. Mr. Hood developed a theory which laid down in tabular form what

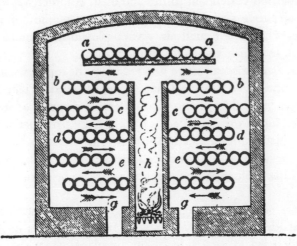

A type of hot-air furnace attempting to admit as much outside air as possible to the ameliorating effect of the fire.
(*The Book of the Garden*, C. M'Intosh, 1853.)

length of flue or pipe how many square inches of fire bar would supply, starting with 100 square inches supplying 200 feet of 4-inch pipe, 266 of 3-inch and 400 of 2-inch.

Hot-air stoves were a direct outcome of the flue system, but as early as 1813 they were being condemned as worse than useless. M'Intosh pointed out that the German (or Dutch) stoves, a hot-air stove, had been banished for a century from British gardens and here were the horticulture theorists attempting to bring them back.

Yet the patents were many and so were the stoves, those most used being Dr. Arnott's, Chunk's, Nott's Vesta, White's, Lawe's, Boyce's, Jack's, Forsythe's, Harper and Joyce's, Deane's, Hazyards and the Polmaise.

Their greatest disadvantage lay in the fact that they heated the air too highly and also that the heat playing upon the metal used in their construction put carbonic acid gas into the atmosphere. Most of the stoves were for placing in the houses without flues of any kind but for the smoke stack which went through the roof. This, of course, meant that firing-up took place inside with all that method's attendant fumes and dust. Another way of using the hot-air stove was that practised in some public buildings of the time, where the heated air was conveyed in channels closed over with iron or brass grilles in the floor.

Allen's Archimedean stove was fed from the top down a central column round which a spiral screw flue revolved to take the hot air from the fire, thus giving a considerable distance of travel for the heated air before it arrived at the chimney. Another stove had the flues placed round a central fire but in zigzag fashion. Some were so designed as to introduce fresh and cold air from outside by pipe into the firebox by means of a cock to be turned on or off at will.

A most elaborate stove, that of Lawe's, introduced cold air from outside the house by means of two short flues on ground level which led to each side of the fireplace where the air then ran into a series of six rows of earthenware pipes 20 inches long and $1\frac{1}{2}$ inches in diameter and was heated by the fire which played all over their outer surfaces before it passed into the house.

There were many more designs, ever so many more, but as soon as the idea of steam and hot-water heating was introduced their death-knell was sounded; indeed many discerning gardeners never used the hot-air stove or even toyed with the idea, but clung to their furnaces and flues until the hot-water-pipe system took over.

There were two quite unusual ideas at the beginning of the nineteenth century for heating, one from Dr. Anderson and the other from Loudon. Dr. Anderson's was typical as showing the temper of the time for experiment, however ludicrous or incredible it might seem. In a treatise of 1803 on a patent hot-house, he had the idea of trapping the sun's heat during the day and conserving it against the cold of the night. The plan showed two greenhouses one on top of the other. The glass laps were to be made most secure to make the house as air-tight as possible and there was to be a minimum of woodwork in it. The general idea was to trap the heat of the sun during the day in the top storey and regulate its entry into the greenhouse below by means of valves. Dr. Anderson stated that the reverse would then happen at night, when the heat in the bottom house could be valved into the upper house to drive out the cold air.

Mr. Loudon's idea of 1805, published under the title of *Several Improvements Recently made in Hot Houses*, was a scheme which, like Dr. Anderson's, came to naught, but was notably worth the effort. This invention used an air-pump to draw the cold air out of the greenhouse and so make room more quickly for hot air to rush in and replace it.

Steam, when it was applied to horticultural heating, caused probably as great a revolution as when it was applied to industry, especially when it led almost directly to hot-water heating, for when both these methods had been tried and many, almost too many for a conscientious gardener's sanity, clever minds had experimented, invented, designed and tested, the way was wide open for the growing of every plant from every clime, every country, from scorching desert, from icy mountain-top, from steaming jungle, snowy fastness and smelly swamp.

But first those early efforts with steam. Sir Hugh Plat, the Elizabethan, as mentioned in earlier chapters, was very much before his time in his experiments with steam, and with his hankering after using the waste heat he could see licking around his stewing mutton or boiling venison for keeping plants warm or for germinating seeds.

But it was not until 1788 that it was applied with any measure of success by Wakefield of Liverpool to hot-house heating. This early method was by means of a single tube leading from the boiler into the house with an ever-present possibility of explosion or scorching. Early amendments were quickly devised, one being to lead a narrow-bore steam tube through the centre of an empty larger-calibre pipe so that the hot air so generated could be directed where it was most needed.

The most descriptive recorded early use of steam was by Mr. Butler, gardener to the Earl of Derby in 1792, who heated his melon and pine-pits at Knowle by the new method. He ran perforated steam pipes under tan-bark beds in which his pots were plunged. The steam was allowed to escape into the tan until it was saturated. Speechly in his Pine Treatise of 1796 gave his plans for steaming pines, but did not agree that steam could be used successfully for heating in a general way. In one plan he gives a design for steam to percolate the tan-beds, in the other he speaks of lead pipes with small-bore perpendicular ones rising from them, all plugged. The boiler had to be heated up to boiling-point, the plugs were then removed from the small-bore pipes and the houses heated with the steam.

In 1807 Mr. Hay, of Edinburgh, used a further improved mode, when he ran a pipe perforated along its length under a mass of rough loose stones three to four feet in depth. The steam was allowed to escape until such time as there was no more condensation on the stones, showing that the mass was as hot as the steam. It was suggested that for a Tropical House the heat so obtained would last for 24 hours in cold weather, or for two or three days in mild. The plants were stood on top of the steamed rubble.

By 1816 steam heating had spread rapidly and big ranges heated by this means were a forcing-house at Kensington Gardens, extensive ranges at Loddiges in Hackney, while Mr. Gray at Hornsey used two boilers to heat 50,000 feet of air in ten large hot-houses, the largest of these 550 feet distant from the boiler. Loudon put forward as the advantages of the system that the tubes near the boiler were never heated beyond 212°, or boiling-point, while at a distance of 2,000 feet almost the same degree of heat was found. Disadvantages were the dry, excessive and metal tainted heat fumes raised by the steam pipes, and the fact, when it was proved, that steam did not, as at first claimed, kill all marauding insects in hot-houses.

As a counter to these criticisms further methods were sought out and used, one being to allow steam into vaults below the floor of the houses to be heated, and another to play that steam on to faggots or on stones or brick rubble in beds. In Bath Mr. Stothert allowed steam to escape into an enclosed flue filled with loose stones until the flue was saturated and the boiler safety-valve lifted. What anxious moments many a gardener must have spent watching this one work! Then there was the tank system in which small-bore steam pipes were run through shallow trays, some 18 inches deep, filled with water. Over this heated cistern was a floor of brick or stone pavement on which the plants were stood.

Probably the most successful application of steam to greenhouse heating was that used by Sturge, near Bath, who heated water in eight-inch-diameter pipes by means of a one-inch steam pipe running through the centre. Through hollow screws in the middle of small shallow trays on top of the pipes expansion water bubbled up for the purpose of furnishing a humid atmosphere in the house.

It was a practice among some to put into the house steam vapour impregnated with nicotine or sulphur to kill insects. In the evaporating pans placed on top of the steam pipes it was also the practice to put unrolled leaves of tobacco as well as guano, pigeon dung and urine as, says a contemporary writer, "the ammoniacal fumes given out by the uses of these latter bring the atmospheres of a Hot House to about the same state as did the old dung beds, known, notwithstanding their dirty appearance and great waste both of labour and manure, to be so pre-eminently valuable for the restoration of sickly plants, the vigorous growth of healthy ones and the total destruction of insects." Is this smelly practice something of which we have lost track, one is prompted to ask, and have our modern glasshouse scientists and research stations proved or disproved the efficacious effect or otherwise of dungy smells on plants? It seems fairly certain that similar practices are not carried out today or we should be hearing, without doubt, of "Smell-money" being demanded by horticultural labourers or of a strike because of smelly conditions in greenhouses.

The system of steam-heating used to heat water pipes led quite logically and naturally to the age of hot-water heating, a practice which by the end of the nineteenth century was practically universal.

The pioneering use of hot water, horticulturally, is credited to the French, and it is a strange and almost incredible fact that it was only a matter of a hundred years ago that it was realised that hot water had circulatory powers. Count Chabanne, a Frenchman, first wrote of this power in 1818, but it was left to the scientists to fight out how it was brought about, and what a fine old fight they had!

The first indication that hot water could be used in a practical way for horticultural heating was recorded by Stuart in *Stuart on the Steam Engine*. He noted that Sir Martin Triewald, a Swede living at Newcastle-on-Tyne, before 1716 had described a scheme for warming a greenhouse by hot water, boiling it outside the house and conducting it by pipes into a chamber under the plants.

Nothing appeared to have come of this scheme, hatched before its time and the next move came from a smallholding outside Paris where in 1778 M. Bonnemain, a physician, used hot water to incubate chickens by a series of hot-water pipes rather like a modern radiator.

Count Chabanne, who must have known of this Paris scheme, first gave an indication in a treatise published in 1818 that hot water circulated, the first time this property had been realised.

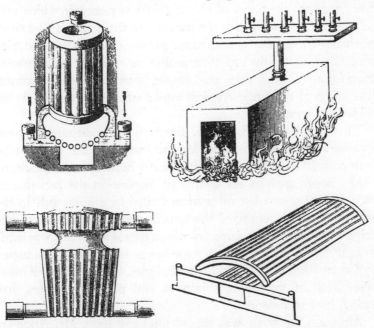

A few of the many "new and improved" hot-water boilers which came on to the market to take the place of hot-air methods of heating. The claim for each and every such was that it presented more water-heating surfaces to the fire than its rivals.
(*The Book of the Garden*, C. M'Intosh, 1853.)

The idea had struck him, he said, when he had been toying with a hot-air stove, that he might as well try water instead of air in his caloriferae, as he called his stove.

"The most perfect description I can give of the circulation of hot water," he explained, "is by comparing the boiler to the human heart and the effect of the caloric upon the liquid to the circulation of the blood in our veins.

"The fire is the power which gives motion to the water, as the admission of oxygen into our lungs causes the circulation of our blood. A pipe is placed at the top which may have any length or winding, but must finally return to the bottom of the boiler. The caloric, which passes into the liquid rises to the upper pipe and communicates itself to the liquid in it, which loses that heat as it flows through the pores of the metal (or a reservoir which may be placed in its passage for the purpose of extracting it), becomes gradually cooler and in that state pressing on the rarefied pipe which comes from the top of the boiler, re-enters at the bottom in proportion to what goes out above, thus causing a continual circulation of the liquid coming into contact with the fire at the colder temperature."

That was Count Chabanne, but two other disputant figures appeared on the British scene, Mr. Atkinson and Mr. Bacon, who both claimed to have invented hot-water heating for horticulture.

Mr. Bacon gave as his source of inspiration the fact that in 1804 he had seen a leg of mutton boiled in a horse-pail in the bottom of which was fixed the muzzle end of a gun, the breech being placed in a nearby fire, by which means the water was made to boil. Mr. Atkinson said an experiment he had seen conducted by Count Rumford, who was the designer, among a host of other inventions, of smokeless chimneys and patent firegrates, had incited his experiments.

Although the battle was waged loud and long, Mr. Atkinson was given the honour of being the first inventor of a practical apparatus, as it seems that Mr. Bacon's system, in his garden at Aberaman in Glamorgan, only consisted of a single pipe some 12 feet long, one end placed in the fire and at the other end a small-diameter pipe which rose perpendicularly some 18 inches for expansion purposes and for supplying water to the system. As was said at the time, when it was examined in November, the result was little or no circulation and an immense waste of fuel, the boiler consuming nearly one-quarter of a ton of coal a day. Much of the water boiled out of the expansion pipe too. An amendment made later by Mr. Bacon was the use of a reservoir instead of the expansion pipe, but he had not in fact realised that water circulated and by sticking to his one pipe, the consequence was failure.

Mr. Atkinson's original system, however, used an open-top boiler with a moveable cover and two pipes connected to the top and bottom of the boiler running to an open-topped reservoir. He so explained his system, and it is an important formula in the understanding of hot-water systems generally: "In order to show the principle of the hot water apparatus," he said, "we may select a simple case of two vessels placed in a horizontal plane with two pipes to connect them, the vessels being open at the top, one pipe at the top and the other at the bottom. If the vessels and pipes are filled with water and heat be applied to vessel A, the effect of heat will expand the water in vessel A and its surface will in consequence rise to a higher level. The density of the fluid in vessel A will also decrease in consequence of its expansion, but since the column of fluid above the centre of the upper pipe is of a greater height than the column of water at the other end of the pipe motion will commence along the upper pipe from A to B and the change this motion produces in the equalising of the fluids will cause a corresponding motion in the bottom pipe from B to A and this motion will continue until the temperature be nearly the same in both vessels, or if the water be made to boil in A it may also be boiled in B because the ebulition in A will assist the motion."

This system circulated water on the level only. By giving the boiler a closed top Mr. Atkinson found he could carry his hot water some 30 feet above the boiler. Then came Mr. Kewley of London, and Mr. Fowler of Devonshire, who discovered that by using the properties of the syphon water could be made to circulate both above and below the boiler and at an increased rate. Hermetically-sealed pipes and boilers came next, the advantages being smaller-bore pipes and higher temperatures.

The reason for hot water circulating still worried the theorists, and Mr. Tredgold, a London engineer, put forward the expansion theory which was shared by Atkinson, Bacon and Barrow, says Loudon. Mr. Hood would have none of this, nor would Mr. Tomlinson who each had different theories. Mr. Tomlinson's idea was that water was not a conductor of heat, but that heated water particles expanded and pressed upon each other ad infinitum to cause circulation. The advocacy of many systems by many men tended to fill the columns of the horticultural journals of the time but while the scientific discussions might well have been followed with avidity, the gardeners were only too pleased to know that water did circulate even if they were not sure how or

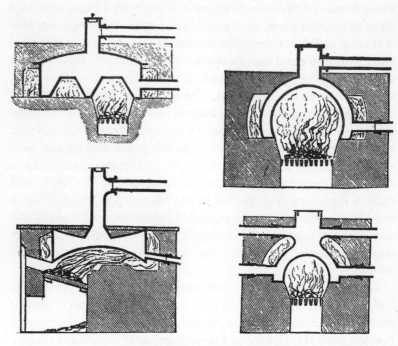

A selection of hot-water boilers illustrating some of the many theories as to the best shape of a boiler for efficient heating.
(*The Book of the Garden*, C. M'Intosh, 1853.)

why. In any case hot-water systems were installed in all parts of the country to replace the hot-air and flue systems.

And hot-water systems—what a proliferation of these there was! At any one time a head gardener might have taken his choice of at least forty different ones, each designed, said their inventors, to be better than the last.

Yet there was still a hankering after some hot air and Mr.

Weeks devised a system in which two-inch pipes from outside went through the hot-water pipes before opening into the house.

Radiators, ornate cast-iron vases filled with hot water from the pipes on which they stood, with lids which could be raised to give humidity when needed, and ornate cast-iron balustrading in which the pilasters and coping were the hot-water pipes, are noted at the time along with heat-boxes, elaborately decorated cast-iron receptacles inside of which were small-scale cellular radiators.

In Penn's system, a peculiar eccentricity, but quite typical of the times in its diversity of invention, the strange idea was conceived of having the hot-water pipes outside the house and "it is

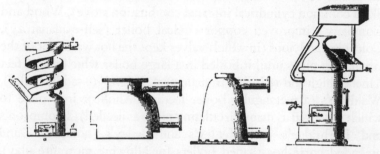

A further collection of hot-water boilers showing ingenuity and straightforward utility.
(*The Book of the Garden*, C. M'Intosh, 1853.)

attempted to bring the hot air in at the top of the house and make it descend." Then there was the Eckstein and Busby method for which the advantage was claimed that the boiler could be placed in a garret and by means of a "propellor or vanes" placed in the chimney and turned by the smoke the water was forced downwards.

As a further indication of the bewildering number of systems M'Intosh lists the following, all differing in some major or minor detail. There was Perkin's high-pressure apparatus (coils and no boiler as such), M. Bonnemain's (small-radius pipes ranged horizontally in several tiers before returning to the open boiler from the top), Watson's (which also heated air and put it through

H

the chimney flue which went round the house before reaching the chimney), Cruikshanks' (free-standing boiler in house), Rogers' (a conical boiler), Rettie's (also an air heater as well), Sampson's (this had spiral flues around the boiler to extract as much heat as possible), Cottam and Hallen's (a horseshoe-shaped boiler resulting in less brickwork casing), Burbridge and Healey's (ribbed surface of the boiler offering most surface to the fire), Garton and Jarvis's (a cylindrical, horizontal boiler with the fire inside and outside it), Garton and Jarvis's double drum (a boiler split in half to catch the last heat of the fire before it reached the flue), Stephenson's double cylindrical (this was fuelled from the top through the centre of the boiler), Stephenson's improved conical (shaped like a common cylindrical internal combustion stove), Wood and Company's improved copper conical boiler (self-explanatory), Coldridge and Sons (in which valves kept the hot water out of the circulating pipes until it boiled in a large boiler which was 6 feet 6 inches high and 18 inches in diameter), Neeve's (a saddle boiler), Waldron's (a rectangular boiler feeding a hollow iron plate to which was fixed as many circulating pipes as needed), Thompson's and Burbridge's and Healey's and Bailey's (horseshoe- and amended horseshoe-shaped boilers for allowing more fire play), William's (with a manhole for cleaning out), Weeks and Company (had upright tubes connected by lower and upper rings with the fire in the centre and hollow furnace bars through which the returning water passed before entering the boiler again), and Kewley's Universal Flue (a boiler in three narrow sections round which the flue traversed). There were others!

8

VENTILATION—ANOTHER PROBLEM

WITH the better glass, better glazing methods, more efficient heating methods and better construction all round of "plant asylums," ventilation became quite a problem. The old idea had been to keep the cold air out, always a ticklish job with the early hot-houses as the gardeners' concern with mats, both inside and outside, and heavy wooden shutters, well showed.

Now air was wanted, regulated air, of course, and again to the rescue came horticultural theorists, engineers and scientists to put forward a complexity of ideas and suggestions.

Dr. Lindley wrote: "When a man builds a Forcing House he settles carefully the slope of the roof, the nature and direction of the heating apparatus, the material for his shelves and floors, the quality of his glass, the size of the squares, and the depth of their laps; whether the door should be at the end or the side, or the whereabouts of the stoke hole, are other points of great deliberation. How then are we to account for the almost universal neglect of that most important part of all—ventilation?"

He added that if he did not hate new words he would be inclined to expunge the word "ventilation" from the language of gardening and substitute for it "zephyration or some other such gentle epithet." As it was, ventilation was defined as letting the exterior air into a house whereas aeration was put forward as the ideal, with a definition of "keeping the atmosphere of a forcing house in motion by currents of warmed fresh air." That was most important, Dr. Lindley stressed, as it was the one thing necessary "to render an artificial climate natural."

The theorists once again fell into the anthropomorphic trap, stating as they did: "A plant condemned to pass its life in a still atmosphere is like nothing so much as a criminal set fast in an ever-lasting pillory." If ventilation meant letting in the wind, then most of the ranges of Dr. Lindley's day were not at all deficient, nor, as had been the theory and practice ever since the first hot-houses, did ventilation mean cooling down a house which was too hot.

"In order to secure motion in the vegetable kingdom," Dr. Lindley commented, "currents of air are made to do the work of muscles, limbs, and volition of animals. It is not at all improbable that in addition to the mechanical effects of motion in assisting the propulsion of sap, it may be important that the stratum of air in contact with the leaves of plants should be incessantly shifted in order to enable them to procure an adequate amount of food, for we find that water in motion feeds them better than that which is stagnant."

When air moved over the surface of leaves, the Doctor explained, minute quantities of carbonic acid gas were extracted and the more air there was passing over the leaves the more food. Moving air, it was said, also assisted considerably the "perspiration" of a plant from its leaves.

It was the new houses, "glass balloons" as some called them, or "mausoleums of horticulture" as did others, which were causing the ventilation troubles. And gardeners "blessed" with the new metallic roofs and sheet-glass glazing were finding their blessed state a curse. Let Mr. Glendinning of the Chiswick Nurseries speak for his fellow-gardeners round about 1850. "Small houses," he remarks, "are easily managed in this respect [ventilation], particularly those constructed of wood and glazed with small glass; with those of great magnitude, of metallic construction and glazed with large glass the case is very different. According to the present defective mode of ventilating, plants seldom succeed so well in such houses as in those of more humble pretentions.

"The bad effects of such houses are soonest manifested in houses constructed of metallic roofs and glazed with sheet glass, thus rendering what is considered elegant and beautiful in these ornamental structures fatal in the application as vegetation thrives ten times more luxuriantly under the huge wooden beams and rafters with green glass (crown) and leaded laps."

Shading was resorted to in the new houses and, as recounted earlier, various coloured glass was used, but many horticultural minds were busy and slowly the problem was resolved. There were many steps along the way, some useful, some nonsensical. Probably the earliest practical and seriously-thought-out system was the use of inch-diameter tin tubes, two feet long, running through the woodwork of the bottom sashes. On the outside of these were tin funnels some seven or eight inches in diameter, and on the inside end were removable fine watering-can-type roses. You could thus have your fresh air sprayed or in a thin stream. For the outside funnel you provided the gardener with corks or plugs and very curious must those early houses (1750/ 1800) have looked with this primitive apparatus sticking out from them.

It was to the well-built lean-to's that the ventilating engineers first gave their attention and much activity and ingenuity were devoted to lifting, sliding and tilting the heavy wooden and metal shutters in both front and back walls. Weird and wonderful arrangements of gears, racks and pinions, worms and crowns, winches, ropes and pulleys, chain weights, handles and extended rods were advertised and used.

Mr. John Williams, of Pitmaston, invented a cylinder in which a piston actuated by the expansion and contraction of the air within operated directly on a sliding sash. By a primitive rule-of-thumb method the machine could be fixed so as to function at certain temperatures, either lowering a sash or raising it.

There were quite distinct differences in the nicety of operation of these gardeners' aids to ventilation ranging from a tolerance of a claimed quarter of a degree right up to 15° before the apparatus would function.

Mr. Muglistone's apparatus also worked on the principle of expanding or contracting air in a closed cylinder, except that the piston in his machine moved a valve which allowed hot air which had collected in a perforated tube, placed either vertically or horizontally inside the house, to escape to the outer air.

Dr. Anderson and Mr. J. Williams used oblong bladders made

fast at one end and attached by means of a cord to a movable pane or small sash at the other. The bladder was filled with air at the common temperature of the house and as the air became heated with the air in the house the bladder assumed a globular form "when its peripheries became closer and pulled down the sash."

Deacon's Eolian machine, Dr. Ure's fan, Dr. Reid's chimney and other strange contraptions were mooted but proved unpractical. One of the fans was to be made as big as the house and rotated under grilles in the floor by human power!

Deacon's machine, another fan type, raised enough wind in a large conservatory to cause the leaves of the trees housed there to tremble and shake. In span houses the ridges were designed to lift bodily, quite a simple but effective system as some of the old houses using it today will exemplify.

Mr. Harwood's system for raising sashes used four bevel gears, crowns and pinions and long shafts worked with handles from the back wall. Many of those machines used brass parts and must have been quite costly to instal.

Many were the designs suggested and used for bringing fresh warm air into houses at floor level. One brought the air through valved gratings outside the house down pipes led over a trough of water warmed by the bottom pipes of the heating system. Another used pipes with stoppable ends, run straight into the house behind the hot-water pipes. Others introduced air pipes actually into the hot-water pipes so that the cold air from outside in its short traverse through a 4-inch pipe was claimed to be warmed before entering the house atmosphere, while those houses with flues ran their cold-air pipes through the flues to warm the contents.

Still another method brought the fresh air from the roof in pipes. With the polyprosopic roofs, a rather complex matter of rods, chain and pulleys opened all the roof glass like a venetian blind. In the span-roofed houses (the real greenhouses we know today) becoming popular at the time (about 1850), thin metal flaps counterbalanced with weights were poised over air drains at the ends of the houses to shut one way keeping out the cold air while letting out the warm. The lower flaps at floor level were arranged to do the opposite.

Fresh air in some houses was brought to the house from below floor level and introduced through the grilles in which ran the hot-water pipes, a very favoured way with the builders of big ranges who quite often ran vaults under the ranges which were built on an elevated terrace.

A costly but neat method of ventilation was the glass panel arranged with strips of moving glass like a miniature venetian blind.

When today we hear of automation applied to greenhouse tending we are apt to think how terribly clever we are getting, but Mr. Keeley as early as 1806 made an Automaton Gardener. This contraption, used by Mr. Keeley first in his own garden at Douglas in the Isle of Man, and later at Colville's Nursery, King's Road, London, would open or shut ventilators, shut or open a steam-valve or damper, and ring an alarm-bell in the gardener's bothy once a set temperature had been exceeded.

Its basic motivation was a thermometer placed horizontally and balanced like the beam of a weigh scale. As the temperature rose or fell, by means of a double tube of mercury and spirits of wine, the thermometer as a whole rose or fell like the beam of a scale and activated wires. These wires worked a centrally poised water-valve communicating with a cistern in which was a float and a piston to the top of which were attached chains running over several pulleys. As the temperature rose so the valve was rotated, water ran into the cistern, forced up the piston, while the chain attached pulled on the sashes and raised them on one side of the apparatus, the chain on the other side activating a steam-valve or pushing in the stove damper. When the temperature fell the valve was turned the other way and water flowed from the cistern to a waste-pipe, the piston fell closing the sashes, opening up the steam-valve and the damper. In the meantime a further wire attached to the "beam" of the thermometer scale would have rung an alarm-bell, if this had been thought necessary.

The cylinder for the machine was not too bulky, 2 feet high and 7 to 14 inches in diameter, but the thermometer must have been a Gargantuan affair, ranging from 2 feet to 12 feet long with the appropriate diameter of thermometer tube, depending on the number of sashes there were to open.

If a gentleman wished to go to the expense he could acquire an artificial rain system in which a leaden pipe of half-inch bore was placed in the ridge of the roof, perforated with fine holes two inches apart. When the water was turned on a fine rain descended on the plants. In 1834 Messrs. Loddiges had such an apparatus to water a house 60 feet long.

9

SHAPE—OBLONG OR ROUND ONES?

THE shape of a greenhouse had given little or no worry to gardeners previous to the nineteenth century. Architects had certainly built many orangeries and some conservatories to match their owner's mansions in Grecian, Gothic, Palladian, Old English and some even in Moorish or Byzantine style. Orangeries were, in the main, flat-roofed with dignified arcaded fronts, and lean-to's had only progressed in minor details from Switzer's glazed walls at Belvoir Castle. Useful amendments and refinements there had certainly been, but they had altered the shape but little. A stout back wall, a roof raked, in the majority of cases, at an angle of 45 degrees and with a short front wall some three feet or so high supporting a perpendicular row of frames or sash windows was the general outline of a greenhouse or stove.

But in the nineteenth century this was no longer good enough for the new gardeners who had realised, as their predecessors had never done before, what a wealth of beauty, of curious foreign and glorious floriferousness awaited efficient housing to make a veritable paradise of any conservatory or indoor garden. They wanted to bring the tropics and the denizens of the jungle to their drawing-room doors, and as gentleman vied with gentleman to put up bigger and better glasshouses, far-seeing men such as Dr. Lindley and Mr. Knight, of the Horticultural Society, burned much midnight oil and wrote thousands of words to the learned Societies formulating houses which would be so shaped and aligned to the horizon as to receive as much as possible of the day's sunshine and light.

Sir George M'Kenzie in 1815 was probably the first to give the problem of greenhouse shape, as distinct from angle of roof fall, serious thought and he set out to find a structure so designed that it would get most of the sun most of the time. This he found by experiment to be a hemispherical figure. He pointed out that the glass in an ordinary forcing-house was set in a plane at right angles, or nearly so, to that of the meridian and more or less inclined to the horizon.

From these facts he deduced that it was evident the rays of the sun would rarely fall in a perpendicular direction on the inclined plane but, at most times, the rays would fall in an inclined

A fussily ornate circular conservatory, including as part of the decoration a central chimney. As much an architect's job as a gardener's with most probably an arched cellar containing boiler and way of entry for the gardener.
(*The Book of the Garden*, C. M'Intosh, 1853.)

direction and were never perpendicular to the plane of the glass. He knew of the bickering between Mr. Knight and the Reverend Wilkinson about roof angles and he was taking no part in that, he explained, as he looked around for a totally different way of dealing with his problem.

He found out eventually that in a dome-shaped house the sun's rays would be perpendicular to some part of it during the whole day of its shining, but, realising that a spherical greenhouse would be both difficult and costly to erect, in practice he advocated a compromise with a quarter of a sphere, a semi-dome placed

against a back wall. The house had not to exceed 15 feet radius
and to be no longer than 30 feet. If it were needed to be longer,
then Sir George advocated that the shape should be spheroidal.

The house was designed to be built of cast-iron ribs, 15 inches
at the base and narrowing towards the apex. Ventilation was not
through the glass at all, but through the back wall, unless—and
here was ingenuity indeed—as the enthusiastic inventor suggested, the dome should be made in two parts and placed on

A beautifully designed "Domical" conservatory of mid-Victorian
times, opulent, spacious, built to impress.
(*The Book of the Garden*, C. M'Intosh, 1853.)

rollers in the manner of an observatory dome when the two
halves could be rolled aside and the contents exposed directly to
the sun and rain.

Mr. Knight suggested that an easier way to build a house and
still retain the curved section was to build it smaller, using a
smaller segment of the globe, and an idea was also put forward,
and quite often put into practice, of building lean-to types of
houses with curved roofs and ends.

Mr. Loudon suggested an acuminated semi-globe. He agreed
that the most perfect elevation was a glazed semi-globe, but to
surmount the difficulty of water lodging on it he advocated that

the globe be brought to a near point. The next best shape, in his opinion, and it went for a lot in his day, was half of this house built against a back wall.

The *ne plus ultra*, according to Loudon, was the polyprosopic house, which had a curvilinear roof surface but "thrown into a number of faces, these all being hinged at the upper angle and all liftable to let in either the sun or the rain."

Ridge-and-furrow roofing invented by Loudon—but claimed as his own idea (he called it Vandyke roofing) by the Reverend Mr. Carlisle in 1828, thus causing another horticultural rumpus— opened out very wide vistas indeed for greenhouse engineers, particularly with regard to the size of houses. With ridge-and- furrow roofing supported inside by iron pillars there was, in theory, no limit to the covered garden area which could be glazed, and indeed houses were built 240 feet long by 220 feet broad and even larger. Some of these houses had tall upright fronts which ran on rails to facilitate the opening of the whole of the frontage to the air.

For those with money and the desire to have novelty and in- dividuality in the greenhouse art there were great domical con- servatories, built cylindrically in shape, as there were square houses, circular and polygonal. There were still being built for special purposes, mainly fruit-forcing, lean-to's, three-quarter span, north light, while conservatories and winter gardens, which reached unprecedented heights of imagination and architectural tastes, in the great walled gardens of the mansions up and down the land, sprouted with pilasters, lanterns, domed roofs, flat roofs, pinnacles and spires like the silhouette of an eastern city.

The time had now really come when opportunity, wealth, desire and the necessary skill were all available and in plenty to bring to English gardens the golden age of glass and floriculture.

10

THE AGE OF GRACE

In Elizabethan times, and for almost a century later, buildings in gardens were intended and used for amusement and for recapturing an indoor atmosphere outdoors, mainly by means of garden houses, banqueting-houses, supper-rooms and summer-houses. Inigo Jones, the architect of the Banqueting Hall in Whitehall, is said to have designed the first building for a garden with his banqueting-room at Beckett in Berkshire.

Up to 1717 the greenhouse was in practice an orangery, and in this country as on the Continent, the terms were synonymous. It was a building with an opaque roof and solid side walls resembling a dwelling-house, very often with rooms over the top for the gardener or for a store. The fronts, facing the south, were normally built with high wide arches, separated by wall piers. Sometimes the spaces between the piers were glazed, and often they were not. Excellent examples are still to be seen, quite a few built by Robert Adam, gracing the surroundings of many stately homes. They are handsome Grecian, Palladian or Roman pavilions in masonry pierced with huge arched windows topped with heavy pediments, cornices, domes or towers; more often than not they were flat-roofed behind their imposing parapets. At Bowood in Wiltshire there is a magnificent Adam example, and others typical of their day and age are at Blickling Hall, Norfolk (1780), Heveringham Hall, Suffolk (built about the same time), at Croome Court in Worcestershire (1760), and at Belton Hall, Grantham.

And that is why in 1712 John James in *Theory and Practice of Gardening* could still speak of greenhouses as large piles of buildings like galleries, which by their fronts added to the beauty of gardens, and were used for the keeping of orange-trees and other plants during winter. He advised that the greenhouses should be so placed that they would serve as a gallery in the summer to walk in when it rained.

Such buildings, used as much for entertainment as for the growth of plants, were intended for quiet philosophical strolls in the winter sunshine among a little greenery and what warmth the arcaded walls afforded.

A change in the use of the garden houses was noted in 1696 by T. Langford in his work *On Fruit Trees*. He wrote: "Greenhouses are of late built as ornaments to gardens (as summer and banqueting houses were formerly) as well as for a conservatory for tender plants, and when the curiosities in the summer time are dispersed in their proper places in the garden the house (being accommodated for that purpose) may serve for an entertaining room."

The Kensington Palace orangery, built for Queen Anne by Wren and Vanbrugh in 1704, was intended equally as a winter promenade garden and Defoe recalls that "the Queen oft was pleased to make the greenhouse, which is very beautiful, her summer supper house." Orangeries indeed did become palatial and grandiose structures as could well be seen at Versailles and in that excellent example (now Number 3 Museum at Kew) which was the orangery built for Augusta, Dowager Princess of Wales, in 1761. It is 174 feet long, 30 feet wide, 25 feet high with quite beautifully proportioned large bow-topped windows, almost from floor to eaves, glazed in casement style. It is slate-roofed and in all a most handsome building.

Yet about the same time Mr. William Belchier's attempts at an actual conservatory building, which communicated with the house as the later conservatories did, were noted at Epsom. Here this pioneer had a grove of some thirty or forty orange- and lemon-trees placed in the open ground. Out of an elegant drawing-room, a contemporary writer noted, through a pair of large glass folding doors it was possible, in winter, to walk through the grove, for the inventive Mr. Belchier had erected a case entirely of glass for the front, the roof and the far end, to put over his orange grove. In this glassy grove, said the writer, it would be prudent to have a movable iron funnel for heat to run against the glass at the end and the front, such as was used in a forcing wall. It was a brave start.

For the first half of the eighteenth century the field was occupied by the earnest botanist with his care only for stoves and the preservation of plant life, and not for amusement, as places for growing flowers or social prestige or as a background to living. Then came the enthusiasts of the late eighteenth and early nineteenth centuries with an eloquence not since paralleled, urging men of wealth, taste and leisure to take up horticulture and the most refined branch of that art, to wit, greenhouse practice. It is quite true to say that there was an amalgamation of circumstances during this era which gave every possible scope to the establishment of the golden age of the glasshouse. There was comparative peace in the world, there was incredible wealth for the few, there was leisure, inclination, and comparative ease of travel, making plant-hunting less arduous and cheaper than it had ever been before. Horticultural science was at its zenith too, and at a pitch which matched superbly in conservation the findings of the travellers and, coincidentally too, British gardeners were more skilled, painstaking and knowledgeable than they had ever been before or since, for what era had ever seen grown with success such an overwhelming wealth of previously totally unknown exotics and rarities from every corner of the known world?

Curiously enough the taste of the nineteenth-century patrons of the art had swung full circle to match the wishes and needs of those early lovers of the garden. Now, as two hundred years before, garden-owners wished for their garden buildings to be a source of amusement, entertainment, a gracious background to their homes, their lives and their way of living. In other words they were back at the banqueting-hall and supper-room stage, but with gorgeously bedecked, light and airy conservatories just on the horizon to take the place of the old, solidly built, opaque-roofed orangeries and galleries.

There were goodly examples of these conservatories and winter gardens to be found at the palaces and country *châteaux* of the numerous Princedoms and Dukedoms in Europe, and up and down the Continent in the great botanical gardens were to be seen tremendous winter gardens and conservatories, and great ranges of fruit- and forcing-houses.

In this country there were many eloquent persuaders abroad who, in gardening literature, loudly and persistently exhorted the man of wealth and leisure to take up greenhouse culture as a bounden duty first, and only secondly as a source of personal pleasure.

On the Continent meanwhile, the English gardener abroad would be able to see the most spacious greenhouse in Europe, the ranges built for the Emperor Francis I at Schoenbrunn between 1753 and 1755. A whole shipload of exotic trees and plants were procured for the first furnishing of this house, being shipped from Martinique to Leghorn from whence they travelled on muleback to Schoenbrunn. Five more ships' cargoes of plants came from Caracas and Havana.

By 1759 the hot-houses had grown to a grandeur appropriate to their wealth of exotic beauty and novelty. One range was 230 feet long and 30 feet high, another 300 feet long and also about 30 feet high, while three more ranges ran to 240 feet long; a glasshouse range 1,250 feet long!

At the end of the eighteenth century Townson, describing these beautiful houses, said "the trees of the tropics develop in full liberty and bear fruit and flowers. The most rare palms, the *Cocos nucifera*, the *Caryota urens*, the *Elaeis guineensis* grow there with vigour, the *Corypha umbraculifera* extends its large leaves for 12 feet round and birds of Africa and America there fly from branch to branch among the trees of their country."

Another striking example was the famous winter gardens at St. Petersburg, described by the Chevalier Storch in 1802 as being built in a great semicircle embracing the end of a saloon, 300 feet long, and lit by immense windows, although the roof was opaque.

Storch said of these gardens, the delight of the populace and of astonishment to all foreign visitors: "As from the size of the roof it could not be supported without pillars they are disguised under the form of palm trees. The heat is maintained by concealed flues placed in the walls and pillars, and even under the earth leaden pipes are arranged, incessantly filled with boiling water. The walks of this garden meander amidst flowering hedges and fruit-bearing shrubs, winding over little hills and producing at every step, fresh occasions for surprise. The genial warmth, the fragrance and brilliant colours of the nobler plants and the voluptuous stillness that prevails in this enchanted spot lull the fancy into sweet romantic dreams; we imagine ourselves in the blooming groves of Italy; while nature, sunk into a deathlike torpor announces the severity of the northern winter through the windows of the pavilion."

Such winter gardens and botanical ranges could be found extending to hundreds of feet and rearing to heights of 60 feet with their domes and pinnacles in many of the principal towns and university centres of Europe. In Paris, which the British traveller had ample opportunity to visit, was the Jardin d'Hiver erected on the Champs-Elysées in 1847. It was 300 feet long and 180 feet wide.

From an elevated promenade it was possible to look down on an English garden with lawns, meandering paths flanked by borders colourfully planted with shrubs and flowers and at the far end a cascade dashing over ornamental rockwork. Fountains threw their filmy delights from the lawns, and in the arms of the corridors—for the garden was designed in the shape of a cross—were planted orange-trees and thousands of camellias.

Principal European cities with great ranges were Brussels, Antwerp, Rouen, Vienna, Baden-Baden, Berlin, Potsdam, Nüremberg, Karlsruhe, St. Petersburg and Moscow. These were magnificent examples which acted as both stimulant and inspiration to the British traveller to return home and do likewise.

Yet it had not been until 1782 that the word conservatory was defined as we now know it, and it was in the *European Magazine* of that year that the *Oxford English Dictionary* notes the following passage: "The idea of a conservatory opening by a folding

I

door into his saloon is too fine to be left unfinished." Sir Walter Scott mentions in *Red Gauntlet*, first published in 1824: "The present proprietor has rendered it [the parlour] more cheerful by opening one end into a small conservatory. I have never seen this before."

In Vienna in 1803 Bory de Saint-Vincent, when he entered the city with Napoleon's army, was quick to notice the new refinement in the homes of the aristocracy there and described the changing scene: "It was a novel and enchanting circumstance, so far as I am concerned, to find the apartments of most ladies adorned with conservatories and perfumed in winter with the pleasantest of flowers. I recall among others, with a kind of intoxicated delight, the boudoir of the Countess of C., whose couch was surrounded with jasmine climbing up daturas set in the open mould; and all this on the first storey. You repaired from it to the sleeping chamber through actual clusters of African heaths, hortensias, camellias—then very little known—and other precious shrubs planted in well-kept borders, which, moreover, were ornamented with violets, crocuses of every colour, hyacinths and other flowers, growing in the green turf. On the opposite side was the bathroom, likewise placed in a conservatory where papyrus and iris grew around the marble basin and the water conduits. The double corridors were not less plentifully garnished with beautiful flowering plants; you might readily, in this enchanted recess, leave open the doors and windows as if an eternal spring had prevailed—the hot-water pipes which promoted and preserved the freshness of the vegetation securing in every department an equality of temperature. Yet all these marvels were kept up at no very great expense."

As was well said in retrospective vein by the editor of *Famous Parks and Gardens* in 1880: "From the beginning of the present century a mania for conservatories has spread contagiously among all the richer classes of the cold or temperate countries of Europe—principally in England and Germany. In Russia, not only the botanical garden, the Royal gardens and the larger private gardens are furnished with them, but even in the towns, as in the towns and villages of Great Britain, they are attached to all the more pretentious houses—their bowers of foliage and blossom frequently communicating with the drawing or sitting room."

At the beginning of the nineteenth century more and more voices rose to tell the rich and influential how they might enrich themselves and the country horticulturally. Loudon, Paxton, Nicol, M'Intosh, and William Cobbett (who was a gardener at Kew in the early days of his chequered career) all wrote in purple prose of the delights of the conservatory and greenhouse. There was Walter Nicol in 1812 to say that the forcing of fruits and flowers and some kinds of culinary vegetables was one of the principal branches of modern gardening. "Since the first introduction of hot-houses into this country," he said, "this branch has made regular progress and has uniformly extended to all parts of the island, in so much as that a garden is not now reckoned complete without a greenhouse, or conservatory with flued walls and with frames and lights."

He alluded, he added, to the gardens of the great and wealthy who found in their glass real pleasure, satisfaction and amusement. He pointed out that they could grow the finest exotic plants and flowers in the world, many of which would otherwise have been known only by their histories. "The botanist and florist have found a fund of constant enjoyment," he said. "The horticulturist never tires, his pursuits are endless. The agriculturist finds here relaxation, amusement and often interest. And all find a garden blooming fair and flourishing at midwinter."

George Johnson, who wrote such a fine history of English Gardens in 1829, added weighty argument to those of his predecessors when he said: "There is not in the whole of the arts and sciences one link in their circle so suitable for the occupation of man in a state of innocency as that which embraces the cultivation of plants and it is an instance of the beneficent providence of the Deity that he assigned a garden as the dwelling of our first created parents."

The great Loudon, writing in 1824, summed up the position

exceedingly well and set the scene for the golden age. "A Green-house," he said, "which fifty years ago was a luxury not often to be met with is now become an appendage to every villa and to many town residences—not indeed one of the first necessities, but one which is felt to be appropriate and highly desirable and which mankind recognises as a mark of elegant and refined enjoy-ment. The taste for these exotic gardens indeed has increased much more rapidly than the skill required to manage them to the best advantage, for the progress of imitation is more rapid than that of knowledge, and hence it is much more common to see a greenhouse than to see one filled with a proper selection of plants in high health and beauty."

Loudon went on to give his ideas on how the greenhouse art began. It is, I think, worth quoting in full; "The most refined enjoyments of society have gradually arisen from desires more simple and even from wants. Man is fond of living beings, and after assembling those plants around him which he found neces-sary for food, he would select such as were agreeable to the eye, or fragrant to the smell. A flower in the open parterre though beautiful and gay, has yet something less endearing, and is less capable of receiving special regard, than a plant in a pot which thus acquires a sort of locomotion and becomes as it were thoroughly domesticated. After choice things were planted in pots, things rare would be planted in them; and from things rare to things foreign and tender—the transition would be natural and easy. Tender rare plants in pots would be taken into the house for shelter and set near the window for light, and hence the origin of the greenhouse.

"In what age of the world, and in what country a greenhouse first appeared it is impossible to determine; it is sufficient to have shown that a taste for this appendage to a dwelling is natural to man; to experience that it adds to his enjoyments and to feel that it bestows a certain claim to distinction on its possessor." So far as the Victorians were concerned, probably the most avid builders of conservatories the world has ever seen or will see, this last "claim to distinction" must have hit right home!

There was also the appeal to the womenfolk by the horticultural writers of the time, and Loudon himself, whose wife later
was to write so many gardening books specially for women, told
the gardening Eves that in a greenhouse was work and interest
eminently suitable for the women of the household. He insisted
that a greenhouse was in a peculiar degree the care of the female
part of the family, for it was recreation for both mother and
daughter in winter when they could not garden outside. "For it is
then," he explained "that the genial climate, the life and growth,
the deep tone of the verdure and the prevailing stillness of repose
within, cause these winter gardens to be felt as a luxuriant con-
secration to man."

It was obvious that even in 1824 Mr. Loudon was using greenhouse and conservatory as interchangeable terms, for when he
later waxed eloquent about greenhouses he stressed that the greenhouse he had been talking about should be adjoining a mansion
or villa. "If it communicates by spacious glass doors and the
parlour is judiciously furnished with mirrors and bulbous flowers
in water glasses, the effect will be greatly heightened and growth,
verdure, gay colours and fragrance, blended with books, sofas
and all the accompaniments of social and polished life."

Even the horticultural engineers emboldened by their early
experiments with hot-water and steam-heating of glasshouses
urged on the populace to build. "The power of imitating other
climates and other seasons than those in which nature affords us,"
wrote Mr. Tredgold, "is known and valued as it ought to be, yet
it remains difficult to imagine the extent to which this power [hot-
water heating] may be applied. In this age it produces luxuries of
which few can enjoy more than the commonest species; but in the
next, nay even in our own, there is every expectation of a considerable addition to the quality and quantity of these artificial
productions as well as to the vast sources of pleasure and information that are afforded to the admirers and students of
nature."

Paxton, building his vast palaces of glass at Chatsworth for the
Duke of Devonshire, was moved, too, by the fascinating garden

vistas he could see. He wanted the Government of his day to take over the gigantic "greenhouse" he built them for the Great Exhibition, his Crystal Palace, as a winter park, a great covered garden for Londoners where he guaranteed he could supply the climate of Southern Italy, where multitudes could ride, walk or sit amidst groves of fragrant trees with creeping plants on pillars and girders.

And, amidst all this encouragement from the horticultural sidelines, garden and mansion owners were doing what they could in the first thirty years of the century when glass was still taxed and bricks and timber were costly. At Carlton House in 1818 there was a conservatory described as, "This unique edifice is composed in the style of ancient English architecture, denominated 'Florid Gothic.' Its proportions and details have been selected with taste and judgment. Its form resembles that of a Cathedral upon a small scale, having a nave and two aisles which are formed by rows of cloistered carved pillars supporting arches from which spring the fans and tracery that form the roof. It is entered by a pair of folding sash doors of plate glass, the interstices of the tracery of the fan vaulted ceiling are perforated and filled with glass, and glass is also in evidence in the pavilioned windows."

This was a first attempt, not altogether successful for the growth of plants. But the introduction of ironwork into glass-house building coincidental with the repeal of the glass tax made light and airy conservatories and greenhouses, which could be formed into the sort of elevation to match the adjoining mansion, a possibility and a challenge to architects and horticultural engineers. John Claudius in his *Hints on the Formation of Gardens and Pleasure Grounds* of 1812, noted copper sash bars and iron rafters "an innovation, the greatest improvement hitherto made in horticultural architecture." He maintained that the principal expense was glass, then tenpence per square foot for third-rate crown.

Humphrey Repton, whose Red Book designs had altered, and were altering, the very English landscape, found time in his major work, *Observations on the Theory and Practice of Landscape Gardening*, of 1803 to spare a chapter on the vogue of the greenhouse and conservatory; but only as it applied to overall design and looked at almost solely from an architect's point of view.

His observations show well enough that the conservatory proper—that is a plant-house attached to the house for beauty's sake—was only just making its tentative appearance. It is perfectly apparent that Repton had no proper conception of what we mean today when we speak of a conservatory, and he, too, as had the gardeners before him, used the words greenhouse and conservatory as synonymous terms. He agreed "that there was no ornament of a flower garden more appropriate than a conservatory or greenhouse; but amongst the refinements of modern luxury may be reckoned that of attaching a greenhouse to some room in the mansion, a fashion with which I have so often been required to comply, that it may not be improper to make ample mention of the various methods by which it has been effected in different places."

He pointed out that at Bowood, Wimpole, Bulstrode, Attingham, Dyrham, Caenwood, Thoresby and other places built in the eighteenth century, greenhouses (he meant orangeries) were added to conceal offices behind them. These pieces of camouflage, said Repton, were designed in the same style of architecture as the mansion, were without glass roofs, and more often than not were only needed to house a few orange-trees and myrtles. The piers between each window were as large as the windows, he complained, and no light was required in the roof.

Now, however, the numerous tribe of geraniums which had been introduced required more light and this had caused "a very material alteration in the construction of the greenhouse, and perhaps, if it more resembles the style of a nurseryman's stove (lean-to style with heat in flues), the better it will be adapted to the purposes of a modern greenhouse."

After conceding so much, the precise mind of the architect took over again to warn that however such an adjoined house might increase the interior comfort of the house, it would never add to the external ornament of it, and it might be better after all to build the greenhouse in the flower garden as near as possible to, but not forming a part of, the mansion. To disguise "the ugly shape of a slanting roof of glass" said the practical architect, treillage ornamentation could be used and still admit light.

Humphrey Repton also put up a conservatory hare which was not caught until Loudon's logic caught up with it some twenty years later. Repton warned his readers that if they joined their houses to their mansions the smell of earth in pots and in borders would more often than not be more powerful than the fragrance of the plants and flowers. Others joined in this heresy and spoke of the danger to health and life itself from noxious gases given off by plant life, the danger of damps and vapours, and the scarcity of oxygen resulting from the plants taking so much for their essential needs. Yet even this was not Repton's greatest objection to the true conservatory and greenhouse; that remained an architect's—that it was almost impossible to make the greenhouse roof conform to the architectural style of the neighbouring mansion "whether it be Grecian or Gothic."

Repton was not the first or the last to make an attempt at these ancient styles and he gives a description of a Gothic conservatory —the two terms are almost at each other's throats—which he designed at Plas-Newyd. This house terminated "a magnificent enfilade through a long line of principal apartments."

The idea, he admitted, was taken from the chapter houses of some of our British cathedrals, where an octagonal roof was supported by a slender central pillar. He suggested that the pillar could be of cast iron, as could the ribs of the roof. If the frames were removable—and this shows up the paucity of the scheme— then it would make "a beautiful pavilion at the season [summer] when the plants being removed, a greenhouse is generally a deserted and unsightly object."

This state of affairs was very quickly to be remedied. For it was not long before there were conservatories which, not only outside but inside, matched the opulence and grandeur of their owners.

1. The Crystal Palace.
2. Palm House at Kew.

(Copyright Country I

3. "The Greenhouse" at Woburn Abbey.
4. Conservatory at Sezincote.

(Copyright Country I

5. Tudor-style Greenhouse at Mamhead, Devonshire.
6. The Orangery at Bowood.

7. Conservatory at Shrubland Park, Suffolk (1830–33).

8. A Victorian Conservatory at Grimston Park, Yorkshire.

9. Nesfield Conservatory at Broughton Hall, Yorkshire.

10. Prototype of a one-piece all-plastic moulded Greenhouse.

11. A large-scale municipal range at Shadwell, Leeds.
12. A modern Greenhouse.

13 & 14. The Climatron—any weather to order.

THE CONSERVATORY ERA

CONSCIOUSLY or subconsciously the tycoons of the early nine-teenth-century industrial revolution must have yearned for some compensating beauty, some exotic means of escape from the dark satanic mills, congested towns and slum backgrounds their very wealth had brought in its train.

For the newly rich with social aspirations without the neces-sary cultural or educational background there had to be some way of doing the correct thing without committing a social or aesthetic gaffe. For while money could buy the country estate or build the pretentious mansion, it could not acquire for the proud owners a knowledge of art, still less aesthetic discretion or taste. But for these social aspirants money could and did buy beauty, colour, art and form in an amazing and inimitable variety by an investment in the products of the natural world. They built con-servatories and gathered about them skilled gardeners to fill them with thousands of exotic flowers and plants and trees.

There was Mr. Dillwyn Llewellyne in Wales who had a con-servatory arranged to represent a tropical forest in miniature. The idea had been suggested to him by his reading a graphic descrip-tion by Schomburg of the falls of the Berbice and the Essequibo in the South Americas. In this house, rockwork was introduced and water, heated to a proper temperature, was made to descend over rocks into a pool which occupied the middle of the house forming an aquarium and a small island of rockwork, which like the rocks round the cascade was covered with exotic ferns, orchids and lycopodiums. Innumerable seedling ferns sprang up among the many specimens of orchids which grew in all their native luxuriance. An innovation, remarked at the time, was the growing of many species of orchid attached to blocks of cork, while others were suspended in baskets hung from the roof, all growing in the wild yet beautiful splendour in which they were found in their native habitat, for the plants were placed in the conditions most natural to them.

In another part of the country let us take a stroll, with a contemporary writer, through the houses and conservatories of a Yorkshire industrialist king of the period, for it gives some little idea of the horticultural splendour which was available to people with money. So we are led into another world from the ugly one outside by the contemporary observer, as he tells us: "A fountain played in the centre of the Camellia House, where camellias were planted in beds, while round the stages were orange trees in pots, cactus and many ferns. There were plants brought from the banks of the Amazon, ferns from Australia, a maple from Japan, a cactus from Bermuda. The Orchid House held many interesting specimens rare in this country. You could have seen peaches, nectarines in profusion in another hot-house; pineapples there were too and a fine banana plant with six rows of bananas. In the vinery were plentiful bunches of ripening grapes.

"In the special span-roofed greenhouse containing cactus curiosities, an African plant suspended from the roof was famous because it required only air for its sustenance, bearing beautiful blue flowers in return. Elsewhere a night-flowering cactus had borne as many as twenty-one blooms. In the delightful fernery trailing plants transformed the walls and roof into a delicate tracery of beauty and the passion flower and climbing plants trained along the roof provided a feast for the eye.

"A rose house produced an abundant variety for your delight, and a rich wealth of colour greeted you in the petunia and geranium greenhouse. Hydrangeas carrying magnificent heads of blue blooms, fuchsias and lemon-scented verbenas shared a house with arum lilies and eucalyptus in good display. Before you left the gardens you would be shown the tulip tree and the magnolia, and the most precious plant you would certainly already have seen, the famous 'filmy fern' from New Zealand, declared to be the only one in England." (Note the sense of competition and proprietorial pride which spurred so many of these wealthy patrons to go one better than their neighbours in the securing of rare and novel plants.) Later this fern was offered to Buckingham Palace and ended up at Kew.

Or for a display of ostentatious wealth combined with a most charming and novel addition to everyday living how about the castle-like conservatory on an estate in Scotland which formed a connection between the house and the garden, matching in its castellated wall the elevation of the garden walls and the house. It was intended to form a pleasant promenade in all weathers for the owners who could not get out in bad weather. A balcony occupied the whole end of the house from which the proprietor, or his lady, could view the whole of the tastefully filled conservatory by stepping out of their private sitting-room.

At Wood End, the Scarborough home of Sir George Sitwell, the very high conservatory full of lofty tree palms and eucalyptus with an aquatic garden beneath, formed a covered way to connect the library and upper rooms with the main part of the mansion, so that the occupants, to get to the upper rooms, appeared to walk on an elevated path resting on the tree-tops.

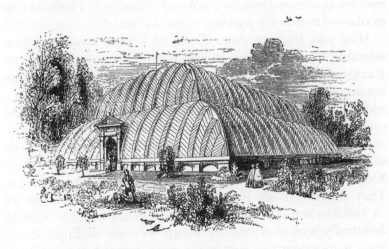

Chatsworth, the Great Conservatory of Paxton, built 1836–40, demolished 1920.
(*European Parks and Gardens*, Anon., 1853.)

The Chatsworth Conservatory—not attached to the house—was a wonder of the age of grace. Designed by Paxton with a ridge-and-furrow roof, this conservatory too, had, at a height of some 25 feet from the ground, a balcony which was carried round the centre part of the house and was reached by a spiral staircase built inside a mass of ornamental rockwork. Inside one felt, it was said, "as if in a new world." Its cost was £11,867, and it took several years to build.

One of its unusual features, later to be copied by its rich imitators, was the divorcement of all utilitarian apparatus from the actual crystal palace itself, so that coal and ashes for the boilers were taken by underground railway from the coal-yard, and the flue from the "subterranean boilers" travelled hundreds of yards underground to a chimney skilfully hidden in the woodland on the hills behind the house.

The Duke himself was highly satisfied with his new possession. "Its success has been complete," he told his sister, "both for the growth of plants and the enjoyment it affords, being, I believe, the only hot-house known to remain in which longer than ten minutes does not produce a state of suffering." There was no smoke or smell in the building either, he added.

Mind you, the expense of all this glasshouse luxury was considerable; in 1846 the cost of coal for the houses totalled £516, £223 for plants, shrubs and seeds, £1,525 for "buildings, hot houses etc.," and Paxton's salary came to £276 8s, to which was added £75 for travelling and petty expenses, bringing the grand total for the gardens and pleasure grounds that year to £9,015 10s.

The seal was set on this spacious crystal case and indirectly on so many of its imitations when Royalty in the persons of Queen Victoria and Prince Albert gave it their unstinted approbation. They drove through the house in a carriage and pair to see "a building and collection of plants so grand and rare as to be deservedly ranked among the minor wonders of England. . . . Such is its extent and convenient arrangement that as many as three or four carriages have been driven in at one time." In 1846 the Duke could write of a total of 48,000 members of the public coming to Chatsworth annually by special railway excursions to a specially built station, to see the horticultural wonder of the age.

Great houses to match, if not in size then in elegance and design, the Chatsworth wonder were now being constructed in all parts of the country. Nesfield at Broughton Hall and Burton at Grimston Hall in Yorkshire constructed beautiful examples of conservatories to match the elegance and architectural style of the mansions of their noble patrons. At Bretton Hall in Yorkshire a most costly lesson was learned by Mrs. Beaumont who for £10,000 built a domical conservatory 100 feet in diameter and 60 feet high. It was the largest of its kind in the world. Messrs. Bailey of Holborn built it of cast iron, it had two huge boilers and was heated by steam, but it was doomed to failure for it was soon found that the height and dimensions generally were too great and no amount of steam-heating could keep this towering edifice warm. Plants were drawn up spindly and weak, and ultimately it was pulled down, the model of it being given by its makers to Queen Victoria as a Wardian case.

One can hardly imagine the Queen paying much attention to that most rebellious of her subjects, William Cobbett, but he echoed her sentiments and put his finger on the Victorian pulse in his *The English Gardener* of 1838 when he urged conservatories and greenhouses on all paterfamilias for the following reasons:

"It is the moral effects naturally attending a greenhouse that I set most value upon," he said. "There must be amusement in every family. Children observe and follow their parents in almost everything. How much better during the long and dreary winter for daughters and even sons, to assist or attend their mother in a greenhouse than to be seated with her at cards or in the blubberings over a stupid novel or at any other amusement than can possibly be conceived! How much more innocent, more pleasant, more free from temptation to evil this amusement than any other? How much more instruction too?

"Bend the twig when young, but here there needs no force, nay not even persuasion. The thing is so pleasing in itself. It so naturally meets the wishes that the taste is fixed at once and it remains to the exclusion of cards and dice to the end of life."

That was a tirade which must have struck right home in thousands of Victorian households where the amusement of mamma and her flock was always something of a problem in an age devoid of mass entertainment, with servants at beck and call and money to burn.

Cobbett added in his practical, down-to-earth manner: "As to the making of greenhouses I shall think of making more than a place to preserve tender plants from the frost in winter and to have hardy flowers during a season of the year when there are no flowers abroad. It is necessary in order to make a greenhouse an agreeable thing that it should be very near to the dwelling house. It is intended for the pleasure, for the rational amusement and occupation of persons who would otherwise be employed in things irrational if not in things mischievous. To have it at a distance from the house would be to render it nearly useless."

In this passage it is most apparent that greenhouse and conservatory had now passed well out of the hands of the professional and the amateur botanist; they were no longer looked upon as horticultural laboratories, as glass quizzing cases, but as part of domestic life and it was this era, this age, which saw in Britain the wonders and the wealth of the horticultural world as they had not been seen either before or since.

Nor was the Queen slow to follow the lead of many of her aristocratic subjects—indeed her advisers must have been some little in advance of public opinion, for at Frogmore, in Windsor Park, she gave orders for one of the most colossal and luxurious ranges of glasshouses to be erected which only that age could have visualised. Its total cost was in the region of £50,000, and building commenced in 1843.

A Parliament survey of 1849 gave an itemised costing list as shown on page 143.

Additions made later brought this royal total to near £50,000. Over 31 acres were enclosed in garden walls, and on a raised ornamental terrace 1,132 feet long was the huge range of glass-

Ground work	£1,000
Garden walls and other general work .	£9,322
Building gardener's house, lime pits, forcing-	
houses, stables and clock . . .	£17,906
Metallic hot-house	£8,921
Hot-water apparatus	£3,908
Tank for water supply	£1,745
Incidental building expenses . . .	£1,028
Plans and superintendence . . .	£1,118

Total £44,948

houses in the lean-to style. Rafters were of iron capped with wood to minimise conduction of both heat and cold, and the sash bars were of copper and hollow, as an aid to non-conduction. Footpaths were of milk-white polished stone, which visitors hardly dare walk upon so virgin was its surface. Hot-water pipes were used for heating, and a refinement was a small-bore perforated cold-water pipe running along the heated one so that water could be sprayed at will on to the hot pipes for the purpose of humidifying the air. There were also 1,665 feet of pits, all heated.

Not long after this splendid example there arose the Palm House at Kew, designed by Decimus Burton and built by Richard Turner between 1842 and 1848. It was, and still is, one of the most inspiring and imposing glasshouses in the world. At the time it caused a gardening nation's heart to swell with justifiable pride and was one more vitally important factor to spur on others into going and doing likewise, albeit on a somewhat smaller scale!

It was a tremendous architectural and building feat—362 feet 6 inches long, the centre being 100 feet wide and 63 feet high, not counting the lantern which soared to another 6 feet. The wings are 50 feet wide and 27 feet high, not counting the lantern and it was all quickly filled with a collection of Australian and Antipodean plants to excite the imagination and whet the appetites of all who saw them.

It was certainly an age for building, for not so far away in Regents Park another glass and iron house arose, again of quite magnificent proportions—235 feet long and 100 feet wide, and some 40 feet high in the central portion, with rockwork introduced as a feature. In the country the Earl of Shrewsbury at Alton Towers had very early in the century, about 1820, put up a most impressive range of architectural conservatories on a great plateau overlooking the Churnet valley, surmounted by seven giant gilded domes of glass. A paved terrace walk hundreds of feet long with a Grecian temple at one end ran beneath, and below that there was a second row of conservatories.

A most unusual feature of the Kew Palm House, as of the Chatsworth Conservatory, was the divorcing of its primary fuelling system from the house itself. In the case of Kew this was most probably done with no other reason than that of amenity, but the idea in other hands became an inherent part of Victorian class-consciousness, like the huge terrace gardens which, overlooked by the principal rooms of a mansion in one of the Home Counties, could only be tended by gardeners in livery, so as not to spoil the view for one thing, and so as not to intrude the working-class type of dress and style upon more delicate and refined creatures within who might be glancing in a romantic reverie over the terrace and parkland beyond.

At Kew, although the boilers, of which there were 12, were in the vaults formed well below the floor of the Palm House, their only communication with the necessary fuel was by a tunnel 550 feet long which also carried the smoke flues to disgorge the smoke via the campanile well out of sight of the house. A small railway ran along the tunnel to the coal-yard for taking fuel in and ashes out. The tunnel though no longer used as a flue is still used for fuel.

Accordingly it was soon after this possible to read in horticultural literature how to make sure that the family should not come into contact with a gardener who might be rather soiled, or might even dare to speak to them, by means of "a private mode of entry to the greenhouse for the gardener."

"If a flower stage occupy the centre of the house," another advised, "a trap door may be placed on the outside where least seen and when the most ready means of entry by a small window in the side of the structure may be effected." One can imagine the scene—abandon ship, her Ladyship approaches—and the gardener and his boy who were busy watering and tending to the plants do a reverse of Venus rising from the waves as they sink

A Paxton patent house, this one built as an orchard-house aviary. The inner arch is a close wire trellis in section, the trees and vines growing on the outside of it. A long list of 23 birds is given as possible inhabitants.
(*The Floral World*, S. Hibberd, 1871.)

through the ferns, or go down like a pair of Mephistopheles in a pantomime.

Now the conservatory was definitely an essential part of the social structure. For dinners, hot-houses, conservatories and greenhouses were raided so that besides the magnificent floral surroundings during the meal, pots of fruit-bearing shrubs or trees with their fruit ripe could be ranged along the centre of the table from which, during dessert, the fruit could be enjoyed by the company; this could have been grapes, peaches, nectarines, oranges, lemons, pears, apples and plums in season, and—this was the criterion of success—well out of season. At Wollaton in

K

Nottinghamshire the staff specialised in taking gooseberries in pots on to the table for dessert.

Sometimes rows of orange-trees or standard peach-trees or cherry, all of them in fruit, surrounded the table of guests—one plant per place exactly behind each chair just leaving room for the servants to approach between. Or sometimes just one well-fruited tree was placed behind the master or mistress who at the appropriate time distributed the golden fruits of bountiful nature.

Drawing-rooms were sometimes laid out like orange groves with sitting-out places made over the pots or large tubs with one or two cages of canary-birds or nightingales distributed among the branches and (spare the head gardener's blushes!) sometimes when it was not a good fruit-bearing year, bought fruit was tied to the tree.

Fruit out of season, flowers out of season, out of their natural habitat, plants probably never seen before in this country, plants brought to light only by the most incredible hardship and journeyings, in other words exotics which needed glass for their growth and protection had become socially important.

Shirley Hibberd, a writer of the mid-nineteenth century, put the conservatory properly into its place in the paraphernalia of civilised Victorian living when he pointed out that the word accurately defined meant an edifice of sufficient size to accommodate camellias and orange-trees and the free movement of full-grown persons attired in a manner which would render it inconvenient for them to come into contact with damp flower-pots. In other words, he added, "a conservatory should be a garden under glass and a place for frequent resort and agreeable assemblage at all seasons and especially at times of festivity."

On these occasions of festivity—a ball, a coming-out party, a coming of age, a wedding—the hall and picture gallery were laid out in imitation parks either in the ancient or modern styles, with avenues or groves or scattered trees. At masques and routs, caves and grottoes were formed under conical stages covered with moss and pots of trees in imitation of wooded hills. Plants of box, laurustinus, laurel, juniper and holly along with pine- and fir-

Another Paxton patent house is designed as a lofty orchard-house by sinking the floor and using 14-foot sashes fixed at an angle of 45 degrees.
(*The Floral World*, S. Hibberd, 1871.)

trees were used to make the groves and woods and the scents of the rose, jasmine and orange were "supplied" for the perfuming of these indoor woodlands.

The ladies had their beautiful corsages of orchids, their stephanotis, their tuberoses; their boudoirs were bedecked with plants and flowers as were their drawing-rooms and halls. In essence, glasshouses provided them with a sweet-smelling, highly-colourful backdrop to their everyday affairs.

The conservatory was now definitely a mark of Victorian pretentions, and again and again the ladies of the household were reminded of this. Their fears were aroused too in this age of the vapours and dizzy spells, for they were warned about gardening outdoors in winter: "What a deprivation of accustomed amusement will a week of snow or frost or of fog, rain or blustering wind occasion [the ladies]. On the other hand what risk of health must be encountered by braving the storm if enthusiasm should supersede prudence." Build a conservatory, a covered garden—

that was the answer, and garden indoors, surround yourself with the horticultural wealth of the Far East and the Antipodes in an English mid-winter; and in towns like Bournemouth, Torquay, Harrogate, Cheltenham and in London town, as can be seen to this day, although now they are soot-encrusted, bleak, dismal, broken-down old places, thousands upon thousands of conservatories were built. Many were most successful and fully answered their purpose, others were vulgar, absurd and could never be used with any degree of success for propagation, cultivation or preservation.

Even the maternal instincts were appealed to and in Loudon's day he tells of a most delightful practice carried out by the "lady amateurs" as he calls them, who, when their geraniums were in flower, would send them to a nurseryman and have them placed where they would be influenced by the genera of other sorts to see what new hybrid or mule would be the result. Other ladies went to considerable expense and trouble to secure nosegays of mixed and desirable-type geraniums and suspended these progenitors over their own plants in flower, hoping again for some unusual offspring to coddle and mother to a colourful maturity.

There were other and more compelling urgent reasons for the building of great houses according to the most efficient and scientific practice, for into the country were pouring all manner of rarities, all manner of horticultural plunder which the plant-hunters were wresting from jungle and mountain side to be seen outside their native habitat for the first time. These were the exotics which needed especial care, special growing conditions and, in some cases, special buildings for their accommodation.

For instance at Chatsworth Paxton designed a special house for the great Amazon water lily, the *Victoria Regia*, which was discovered on 1 January, 1833, and in 1850 was brought to Chatsworth to find a home in the Paxtonian crystal case, a structure $61\frac{1}{2}$ feet long by 46 feet wide and 22 feet high, to cover the central circular water tank 33 feet in diameter. And what joy when this marvel of the Americas actually bloomed! Paxton beat Kew to the blossoming time by a short bloom and he wrote to the Duke, then in Ireland: "Victoria has shown flower. An enormous bud like a poppy head made its appearance yesterday. It looks like a large peach placed in a cup. No words can describe the grandeur and beauty of the plant." To Mr. Hooker he said, "The sight is worth a journey of a thousand miles."

A flower and leaf were carried hurriedly to the Queen at Windsor, the Duke hastened home from Ireland. When at last it did flower at Kew, thousands of people flocked to see this wonder of the age. There it was in its special house in a humid jungle temperature of 85° F. At Chatsworth it bloomed each day at 2 p.m. but by ten o'clock the day's flower had begun to close and all next day died a lingering death, draped, as it was described at the time, in "Tyrian splendour," finally vanishing beneath the water from which it had sprung.

At Oxford, too, the authorities had to keep up with the horti-cultural discoveries and a stove was built there in 1835, said to have housed in 1837 Kewley's siphonic hot-water system for the first time; some years later in 1851 the cult for aquatic plants, culled maybe from oasis side or rushing streams in all parts of the world, had made a new type of house a necessity. Two stoves also were a requisite for the ferns and orchids which came pouring in needing shelter and warmth. The *Victoria Regia* was also housed at Oxford in a tank with an overshot wheel to keep the tepid waters in gentle motion.

Loudon, some twenty years before, had waxed wildly enthusi-astic over the advantages of glass in its application to gardening and gardens, and while the great conservatory at Chatsworth and the Palm House at Kew and some of the other giant houses on estates up and down the country were certainly ambitious, if not grandiose projects, they did not quite live up to Loudon's rosy dreams of glass-covered towns warmed by hot-water pipes, of hot-water lakes and glass-roofed conservatories 150 feet high, so that the tallest forest trees could grow undisturbed and flights of the appropriate birds could fly among their branches; the covered garden *in excelsis*.

Monkeys, too, he mused, could well be introduced to these

jungles under glass as well as oriental birds and other animals, with tropical fish besporting in the heated pools fed by running streams made to operate by machinery. In a salt lake, he added, the fish of the sea, polypes, corals and other denizens of the deep could be kept, or in a freshwater lake freshwater plants and fishes could be introduced. Certainly the Victorians had their small pools and ornamental fountains, and the bigger ones some sort of waterfall, but alas, Mr. Loudon's dreams were just a little too enthusiastic.

Yet Kew's Temperate House, while not quite up to the Loudon standard, was in every way a most noble and magnificent effort on the part of the authorities, the glasshouse engineers and the architects. First started in 1862, it was not completed until 1894 at a total cost of £43,000, which is a lot of money even today. It covered one and three-quarter acres and was claimed then as one of the glasshouse wonders of the world.

Decimus Burton designed this beautiful house, as he did the Palm House, with its total length of 528 feet, to house the sub-tropic splendours, the tree-ferns from New Zealand and Australia and the new wealth from the Himalayas, China, Chile and Japan, the lapagerias, the lily of the valley tree (*Clethra arborea*), the luxurious camellias, oranges and lemons, tender graceful conifers, and the rhododendrons from the Himalayan slopes. From the gallery surrounding the central house there was a bird's-eye view over the tops of noble tree-ferns, palms, araucarias, the bunya-bunya, the Norfolk Island pines, the bananas, the wattles and all their associated flora, representing a then strange and almost unknown world. It was claimed that the visitor had the finest spectacle of its kind to be seen in any European botanical garden.

Naturally in an era noted for its display of wealth Kew's "glassly opulence" was not an isolated instance. At Woburn the Duke of Bedford had long ranges of hot-houses in the middle of which was an apartment fitted up for the entertainment of company in the height of the fruit season. The ceiling was colourfully and unusually decorated with paintings of various kinds of birds, and on the walls hung still-life portraits of fruit by an eminent artist of the day.

At Sion House on the Thames-side opposite Kew, the Duke of Northumberland had a princely structure with a dome like St. Peter's, some 60 feet high, and a long architectural conservatory strung out on both sides of it in a great shallow semi-circle designed in the Italian fashion and terminated with two glass pavilions. The whole of this magnificent glass, 280 feet long, was built on "a chaste architectural terrace."

But the cult, craze, call it what you will, in the Victorian era to surround oneself at all times with as much of the horticultural wealth, colour and rarity of the world as was possible, was not now the monopoly of only the great of the land. Their imitators, nay their rivals, were widespread; the lesser gentry were as avid for botanical novelty, for the possession of rich kaleidoscopic floral and foliar displays as any Duke in the land. They strove with might, main, money, and with every gardener at their command to outdo their neighbours in rarity, to outpaint him in colour, to outstrip him in novelty and to bewitch him with a floriferous splendour one hoped he could not surpass.

An exacting call upon all the greenhouse gardener's art and skill in Victorian and Edwardian times were those fragrant and vividly coloured displays which almost always graced the great hall or ballroom of the houses of the rich at all times, but were particularly rich and opulent at times of festivity.

From a fresh-as-dew lawn of the greenest turf, grown in seed-boxes for the occasion and laid over protective canvas, grew a spring garden in mid-winter. An old tree-stump ivy draped, artfully placed so that it looked as if it had never had any other home, very often made a central feature, and playing underneath its protective branches were to be found a squirrel or a rabbit (lifelike examples of the taxidermist's art). Round the irregularly shaped lawn would be laid out a complete garden in a riot of colour, edged with moss-covered tree trunks. From pots nestled far underneath a pleasing surface of bracken, mosses and fallen leaves, on the lawn edges would be great fragrant drifts of hyacinths, narcissi, daffodils, and on the outer rim of this indoor garden the motif was repeated with begonias, primulas, huge banks of hydrangeas, funkias, and astilbes. Most skilfully placed amidst the blooms were beautifully floriferous examples of the forcer's art, and weeks before their time would be blooming the cherries, the laburnums, the lilacs, forsythias, pastel-leaved acers, the sweet-smelling azaleas and the rich glowing candelabras of the rhodo-dendrons. As a sombre background and contrast two or three dark, dignified fir-trees would have been carefully lifted from the woodland nursery to grace the scene and add reality to the picture. What matter that the display required thousands of plants, three or four days' work for many men to prepare the garden for a few hours' entertainment, and for months beforehand the unremitting attention and most precise timing on the part of the gardener. That was what gardeners and greenhouses were for!

Another beautiful and striking use for greenhouse and con-servatory plants was for what was called the floral table. Here the dining table was arranged as a hollow square or circle with the guests sitting around the outer edge from where they then feasted their eyes through the meal on a sumptuous floral palette. More often than not this was put on in the winter when the stove-house foliage plants really came into their own for a brief hour of glory.

It was usual to centre the display round a fresh green lawn from grass specially grown for the occasion. Quite often a small pond would fill the centre, overhung with all manner of ferns and mosses and leading from it would be a meandering path covered with autumn leaves and from a similar carpet of leaves there were then placed all around, as naturally as art, skill and taste could make such artificiality, hundreds of the taller, many-coloured crotons, the dracaenae or dragon-trees, the bizarre an-thuriums in variety, stately dieffenbachias, the great heart-shaped ornamental leaves of the caladiums, the distinct and unusually marked marantas, the swordlike zebra-striped eulalias, and the variegated abutilons. Then, to add lightness and even more brilliant colour to those of the foliage plants were orchids, cyclamen, *primula obconicas* and *malacoides* in variety and chrysanthemums in bewitching array.

Each and every day of the year from the gardens went the head gardener to see the master or the mistress or both, to learn the needs and the desires of the household. Already a small army of gardeners had been up with the early dusting servants to refurbish the conservatories, refill the numerous vases in all the principal rooms, rearrange their floral and foliage displays in hall and dining-room and, later, see the appropriate floral decorations were in every bedroom and my lady's boudoir. Long before the master and mistress had seen the light of day the conservatories were tended and watered. Broken, dry or dead branches must be removed, new plants taken in and old ones taken out, floors mopped down, glass polished, stagings tidied so that at no time did the indoor garden look the worse for wear for one minute. So far as its owner was concerned Nature never died in the conservatory and flowers perpetually bloomed there.

There were definite cults too in the collecting of plant families. At Woburn the Duke of Bedford claimed to have 800 species and varieties of heather and at Claremont a famous collection of cactus and succulents totalled 900 species, while others specialised in geraniums, carnations, the forcing of exotic fruits, or the collection of all manner of economic plants such as the coffee, cocoa, arrowroot, ginger and so on.

But apart from the specialities and the specialists, be they never so enthusiastic and proud of their collections, it was an age of the all-rounder, of great ranges of glass in which a host of gardeners grew almost everything which could be grown. By 1885 Thomas Baines could write of British gardeners that they were far in advance of those of any other country because of the superiority they had attained in the individual culture of the immense number of species and varieties they grew.

Private gardens were maintained with princely liberality, expense was no object. It was true to say that no gentleman's garden establishment, worthy of the name of gentleman or establishment, was complete without a plant-stove, a pine-stove, its vinery, heathery, orangery (very often a heirloom from the estate's past), and houses for tropical plants, ferns, camellias, carnations, both perpetual and malmaisons in separate houses, geraniums, bulbs, succulents, an aquarium, a conservatory, and, on the utility side, houses for the forcing of peaches, cherries, figs, plums, apricots, pears and the vine.

The fashion for carpet bedding and bedding-out generally on the terraces, lawns and the old parterres of the stately homes called for a tremendous amount of greenhouse work in propagation, pricking-out, boxing, and the subsequent tending and handling from house to frame and final bed. Some authorities credit Paxton with creating the craze the Victorians so obviously had for bedding, but whether that be so or not, the climate and conditions of the times were ripe for such a colourful practice.

It had been for many years—over a hundred—as we have seen, the general practice to stand out in their pots by walks and paths the contents of all greenhouses in the summer, as it had been before that time the practice to fill the parterres with miniature evergreens like box and myrtle, coloured gravels and have an orange grove nearby. So that when more efficient heating and better-designed houses were available to gardeners bedding-out was almost a natural concomitant along with the greater choice in plants and the greater ease in raising and handling.

There were too the Victorian social divisions of the year, as strict as the laws of the Medes and Persians, which meant that when the town houses were closed down for the summer, the masters, mistresses, the family and their guests departed to the country, to descend upon a rural scene in absolute contrast to the city. They had to have laid before them flowers in quantity, in colour and in variety and in great but tidy masses as pleasing to the eye as they were to the nose. It was not unusual for a head gardener and his staff on a big country-house estate to turn out of his houses and frames upwards of 50,000 bedding plants a year.

As for carpet bedding proper, which matched in plants the old geometrical beds so intricately laced and knotted in coloured soils, stones and chippings with but box edgings and centre pieces, these intricate works of the gardening artist used up astronomical numbers of tiny seedling plants. One such ambitious scheme at Cleveland House, Clapham, in 1876, called for no fewer than 60,000 plants.

There were head gardeners in those days, aristocrats of the profession, with three thousand foot run of glass under their care, which needed, and got, unremitting attention as well as unlimited fuel both day and night.

These were the days of the bothy system, where the large staffs of unmarried gardeners, the apprentices and young journeymen, lived communally in the bothy in the garden, or the gardeners' mess. Days were from six to six, with night-shifts in winter for firing and occasional weekends free from Saturday tea-time. Normally the gardeners took it in turn to cook for their colleagues and food was bought on the mess system. Old gardeners who still remember the system speak highly of the "family" spirit on the old estates. "We were one of the family," they say, "and the master and mistress were always careful of our welfare and comfort when we were young 'uns."

There were outside foremen and staffs and there were inside foremen and their staffs, and let it be said now that a fully experienced inside man, often a specialist at his job, be it orchids, inside fruit, succulents, or tropical plants, was a man apart; he knew it, he showed it, the others knew it, and he was able to keep his proper place in the closely regulated hierarchy. Why, an orchid man would not even soil his hands with a sweeping-brush, and a spade was to him a foreign object!

More often than not he was a confidant of his employer and enjoyed a garden classlessness quite foreign to the Victorian scene. It was a world made for head gardeners, who were kings in their own right, like Paxton, who was not only knighted but was a public figure and treated as on equal terms by the Bachelor Duke.

Among head gardeners there was a close freemasonry which saw to it that good keen young gardeners were superbly trained and passed on between the big estates on a head gardener's recommendation only. It was recognised that a young man should move and that head gardeners with specialities would initiate the novice into the mysteries of his particular art before passing him on to another head gardener colleague who had a different but just as important speciality. As many as fifty were on the garden staffs of the big houses and it was quite common with a good glass range to have twenty-five. A typical garden establishment of the sort run by a well-to-do professional man with a county background would have some twenty different houses, and if the master were devoted to orchids he would think nothing of putting some eight of them down to his favourite flower and paying as much as £1,000 for an unusual specimen and amassing a collection running to some 18,000 plants.

Yet despite any speciality the whole establishment had still to reflect in the highest degree the opulence, impeccable taste and fastidiousness of its owner. The stove-houses and conservatories had to be magnificent show-houses where guests could be suitably impressed with rare and exotic plants and high colour at all seasons. On the fruit-growing side peaches had to be available in perfect condition from May to September and always kept carefully in the immaculate fruit-houses for a fortnight or so to secure just the right aroma, flavour and bloom. Grapes had to be available from the fruit-house for Christmas and woe betide the gardener who sent an imperfect bunch up to the house; bunches had to be preserved and so carefully kept, tended and turned in the innumerable grape bottles at an angle on the shelves, that each bunch was perfectly symmetrical in shape, the individual fruit of equal size and all with a perfect bloom on their faces. At Welbeck Abbey, home of the Duke of Portland, 10,000 strawberries in pots were forced annually.

Guests down from town always had to have a hamper of the best fruit to take back, and during the season when the family were in London as many as four great hampers a week of fruit, vegetables and flowers were despatched from the gardens for the delight and delectation of the master. Only the best of everything, and only fruit, vegetables and flowers in the absolute "green" of condition ever saw the inside of those baskets.

Glasshouse gardening in this era was nothing if not exacting and it called for a monklike devotion to the job and an enthusiasm for perfection shared by both master and man. Nothing was too much trouble and not so much as a dead leaf, a broken twig, and certainly never a dead insect—goodness knows what would have happened had there been a live one—was allowed to mar the ordered beauty of a Victorian conservatory.

By the late nineteenth century there was no doubt whatsoever as to the respective roles of glass structures, and in a score of books one could have defined meticulously the differing roles of the greenhouse, the stove and the conservatory and found innumerable designs and details for the construction of every possible type of house.

It was the era *par excellence* of the conservatory, as we have sought to show. Whether their owners used these gardens under glass as a decorative, cosy repository for plants brought there in their flowering seasons from other houses, or whether they were laid out in beds with luxurious palms, sub-tropical trees and climbers, introducing rockwork, fountains and pools, it was a place of resort for the whole family and as much a part of the mansion as the drawing-room or billiards-room. Indeed, according to the Victorian novelists, and those who immediately followed, more proposals must have been made and accepted or spurned in conservatories than in any other locale of the Victorian domestic scene.

The demand for newer and better novelties, for rarities from far-off lands, to fill these ornate and quite often magnificent adjuncts to a Victorian home, was never more insistent and at any one time hordes of plant-hunters employed by the great nursery firms of this country and by horticultural societies and principal patrons must have scoured the jungles and mountain terraces of the world. There were an almost incredible number of plant-hunting martyrs and an epic is still to be written of the journeys, incredible hardships and amazing adventures of these intrepid men who sought to fill the conservatories of the Victorian rich with newer and ever rarer plants.

It became as much a matter of honour to have a conservatory stocked with the newest and rarest exotics as it was to keep the best table and have the best horses in the stables.

Nothing less than the best and damn the cost was a Victorian maxim for many. Thousands of pounds changed hands for the stocking of glass ranges at all seasons. The British led the world as superb greenhouse horticulturists and the one-time Continental leaders were left well behind.

To the following chapters we leave the details of the plants they grew, how the conservatory and greenhouse scene altered completely in a matter of fifty years, how hardly the new rarities were won from their native shores, and how the overwhelming demand for jungle, mountain and forest beauty produced a new occupation, that of plant-hunting.

12

FLORAL TRAVELLERS

Apart from the obvious reasons—poor heating techniques, wrongly designed houses, lack of travel facilities—one major reason for limiting the scope of greenhouse flora was the extreme difficulty of keeping plants alive and seed viable on the incredibly long journeys in the days of pack-mule, sail, horse and carriage.

At first, therefore, only the hardiest of plants from nearby Europe reached these shores, and, as we have seen in previous chapters, the first horticultural imports were orange- and lemon-trees, myrtles and some succulents. Of course there was no great organised nursery trade as we know it and all through the early gardening days of the seventeenth and eighteenth centuries commerce in plants was almost non-existent, except on a friendly basis between noble lords, embassies and consulates abroad, and the very few people who could afford foreign travel and who had relatives and friends in this country; with the probable exception of Peter Collinson, the London merchant, of Mill Hill, who inspired John Bartram to botanise and seek out the curiosities of both flora and fauna in North America for rich British patrons from 1735 to Collinson's death in 1768.

The records would seem to point to the Earl of Portland as one of the earliest plant-importers, for in 1696 he was so enamoured of exotics that he sent his man Jacob Reede to the West Indies to collect new plants for the Royal Garden at Hampton Court. But all gardening history has ever recorded as unusual in the early days at Hampton was a "black" maidenhair fern.

By the middle of the eighteenth century the monotonous attraction of the citrus had disappeared and, in the words of Linnaeus, the "Golden Age" of botany was born.

Sir Hans Sloane heralded the era as an indefatigable traveller and plant-collector, followed in the first ten years of the eighteenth century by Sir Arthur Rawdon of Moira, Ireland, who sent his gardener James Harlow to Jamaica. This conscientious searcher after rarities is recorded as having returned with almost a shipload of plants, which found their way eventually to the greenhouses of the Bishop of London at Fulham, of Dr. Uvedale of Enfield, and those of the Duchess of Beaufort at Badminton. Although Pulteney mentions that among the favourable circumstances which contributed to the golden age was the greater taste for the cultivation of exotics which sprang up among the great and opulent after the happy return of internal peace, he mentions also that the greatest obstacle to securing a really representative collection of exotic plants was the difficulty of transport.

All manner of devices were invented and suggested to help the plant-collectors to preserve their finds, only collected after the most hazardous and dangerous travels. It was advocated that bags made of old tarred rope were excellent for the transport of seeds, whilst others suggested seeds should be packed in charcoal, in closely corked bottles, in sugar or embedded in various gums, wax or tallow. Bartram, collecting seed in the virgin forests of North America, was advised by Collinson to get animal bladders, cut off the necks, put the newly found plants with a little earth in the bladders, add a little water, then tie up the necks of the bladders round the plant stalks, leaving the leaves and flowers outside, and to tie the whole lot to his saddle pommel. For Collinson's precious seeds, paper almost as precious was sent from London, from which Bartram made little bags. Most plants were carried by Bartram in two wicker baskets secured pannier-like on his horse.

When travelling by sea, it was advised that seeds in bags should be suspended from the cabin rafters in an airy position. Other methods were to wrap the seeds in paper and then put them in glass, earthenware or stone pots, covering the stopper with pitch or resin before putting the bottles in boxes filled with sand and placing the boxes in an airy part of the ship.

A detailed account of early methods of seed and plant importation is given by Dr. John Coakley Lettsom, who drew up his instructions with the assistance and advice of Dr. John Fothergill, of Upton, who, in the last years of the eighteenth century—he died in 1780—had probably the finest collection of exotics in the country.

Seeds, and there is little doubt that the great majority of this traffic was in seed and not in living plants, of the larger varieties such as mango and the nut family, could be preserved, said the writer, by rolling each seed in a coat of beeswax about half an inch thick, then putting a number of them so prepared in a "chip box" filled with melted beeswax. The outside of the box was to be coated with a solution of sublimate of mercury and then the box had to be placed on shipboard in a cool and airy place.

For smaller seeds it was advocated these should be wrapped in paper or cotton which had been steeped in melted beeswax, and then placed in layers in a chip box filled with melted beeswax. "Mimosa, japonica and *Aeschynomene movens* from the East Indies" had been sent this way with success.

Another method which Dr. Fothergill had tried and found successful was that of putting small dried seed into cerate paper or cotton and packing them in glass bottles well corked and covered again with a bladder or leather. The bottles were then to be placed in a keg or box filled with four parts of common salt, two of saltpetre and one part of sal ammoniac, "in order to keep the seeds cool and preserve their vegetative power."

Slightly less troublesome was another method of putting the seeds in linen or writing paper and putting the parcels into canisters, earthen jars, snuff-boxes or glass bottles. The interstices between the parcels, it was advised, should be filled with whole rice, millet, panic, wheat bran or ground Indian corn well dried. To prevent injury from insects a little camphor, sulphur or tobacco had to be put into the top of each vessel which was then well sealed to "exclude the admission of the external air."

Some seeds could be placed on layers of moss in a box and

L

allowed to germinate on the voyage home. Bits of broken glass, it was suggested, should be mixed with the moss and earth to discourage rats and mice.

For the few plants which could be carried over the seas with safety, Dr. Fothergill and Dr. Lettsom advocated that a box four feet long, two broad and two deep should be made, hooped with cane, netted, and canvas fitted to pull over the netted hoops. When "white caps appear on the waves" the Captain had to be instructed carefully to pull the canvas blind over the plants or if (and what an "if" this must have been in those exacting days of sail) the Captain had not noticed the white caps and seen that spray was dashing on the plants "in the minutest droplets" then he was expected to drench the plants all over with fresh water!

Even better, if you could persuade the Captain, said Dr. Lettsom, the plant boxes were best fixed up in the "great cabin" near to the stern windows where they could be given air whenever the weather was propitious.

Other boxes for plants were made especially stout with slatted fronts to open and shut, and these, it was advised, should be lashed to the mast on the poop deck, while sailors and officers had to be "sweetened" so they would see to it that the boxes were opened to the sun on fine days, shut down and tarpaulined over in stormy weather, and moistened with fresh water when need be. The tops of the boxes, in these early days, were sometimes covered with talc, Chinese oyster shell or thick glass. Yet even with accommodating Captains, and the kindest of weather, rats and other vermin often left but skeletons of plants by the time they arrived in London River after journeys lasting months, and these were all the fruits of thousands of miles of travel in inhospitable country.

Another method with plants was to wrap them in clean paper separately, place them in clean straw in a close net and suspend them from the cabin roof where they could be kept under the eye of the traveller himself. Earlier, Evelyn had told the Earl of Sandwich in 1668 to try putting plants in barrels, the better to stand long, rough sea journeys.

The wonder is not that we grew the plants in our greenhouses but that they got here at all, and the truth of the matter is, of course, that not many did. It was roughly calculated by Livingstone in the H.S. *Transactions* that one plant in a thousand survived the voyage to this country from the Far East.

Just think of the climatic differences under which many of the plants the collectors found had to live. It is recorded that plants can and do live in a range of temperatures from 30° to 40° below freezing-point up to 170° or 180° F., in light conditions which vary from high noon at the equator right down to meagre candle-light under a forest canopy, and atmospheric conditions from arid desert to steam- and rain-sodden tropical jungle.

In view of all these difficulties then, with what were the greenhouses of the botanists of the late seventeenth and the eighteenth century filled? Earlier, as we have seen, citrus plants, olives, myrtles, and in 1645 Oxford Physic Garden could boast of bamboos and sensitive-plants. John Parkinson in his *Paradisus Terrestris* of 1629 and his *Theatrum Botanicum* of 1640 does not mention greenhouse plants at all although he was growing plants from Virginia, the North coast of Africa and some from Greece and Asia.

Previous to 1700, says Johnson, exotics cultivated in the glasshouse establishments of the nobility did not exceed 1,000 and these were but trees and shrubs like the acacias, myrtles, oranges and lemons, several sorts of jasmine, some pelargoniums, aloes, an assortment of succulents, some economic plants such as tea, coffee, sago and the "fine greens"—a phrase with which the early horticultural writers so often clothed such greenhouse subjects as box, bayes and herbs. It is true to say that "exotics" was rather a high-sounding phrase for most of these inhabitants under glass.

During the century, however, the picture changed altogether and by the end of it 5,000 new exotics had been introduced by the enthusiastic gardeners who wrote hither and thither, nagging their friends and using their influence with Government employees who were carrying out consular or ambassadorial duties in other lands, for seeds and plants. Sea captains were button-holed by wealthy merchants like Peter Collinson, and no traveller to foreign shores would have ever dreamed of arriving back without some choice seed or plant for either his own gardens or those of his relatives or friends.

Bradley in 1718 gives a list of plants for over-wintering—and that is as yet all the glass structures were used for—for all these plants had to stand our summers out of doors. He gives five types of myrtle including the orange-leaved, the Spanish and the Portugal, jasmine (Spanish and Indian, Portuguese and Arabian), the coffee-tree and geraniums including the round mallow-leaved, the columbine-leaved, the spotted ivy-leaved one with a smell like rhubarb, and a large bunch of purple flowers, melianthus, olean-der, malabar-nut, Leonotis, Amomum plini, solanum, caper, Canna indica, genista, date-palm, aloes, torch-thistle, melon-thistle, and ficoides in fifty sorts.

Some idea of the look of these early collections can be gauged by the flower list which Bradley gives; so that in January those early enthusiasts could see under their glass in flower the aloes, ficoides and the Indian and white Spanish jasmine. In February they had the pleasure of some geraniums and sempervivums, and by March there was the Indian yellow jasmine, fig marigolds and some more aloes. Come April there were to be seen some orange-blossom and, making brave attempts under difficult conditions, ficoides, aloes and some geraniums. By May, and for the rest of June, July and August, until the second week of September, the greenhouses were cleared and all plants stood out in the garden.

Not a particularly varied picture for either the botanist or the flower-lover, was it? Why, in 1729 when an aloe (*agave sobo-zifera*) flowered in the Hoxton garden of Mr. Cowell, people flocked from far and near to see it. There was great discussion as to how to preserve the plant in winter and a brick case with glass at the front and an open top was raised around it up to a height of 20 feet. The top of the case was kept open until the flower came into full bloom, when it was closed in.

So great was the excitement that an incident of eighteenth-century hooliganism wove itself around the event and was related by Mr. Cowell. "When the aloe was in great perfection three men habited like gentlemen started to tear off the buds, and when I remonstrated to save my plant from injury, immediately one who was on the top of the staircase in my aloe house being entreated to come down fell aswearing and drew his sword upon my man, telling him he would run him through the body if he offered to assist me and in the meantime kicked me on the head while I offered to go up. While ascending at the bottom of the stairs, one of his companions pulled me by the legs, and the third of them wounded me with his sword in two places in my neck, so that I was under the surgeon's hands many weeks."

It was the same Mr. Cowell who told of the first aloe having come to this country with the oranges to the Carews where they were over-wintered in a pit covered with boards, and that the one that flowered with him had cost his business predecessor, thirty-six years before, 500 guilders.

At that time—1732—when Mr. Cowell wrote his *Curious and Profitable Gardener* he listed the torch-thistle as the favourite exotic of the day, which he said "would stand abroad from the middle of May to the middle of September and then need no other shelter but a common greenhouse where frost cannot enter."

He advised gentlemen with hot-houses that the banana was an excellent subject—"one of the most beautiful plants in the world"—but that as the plants grew high, some 14 feet, the stove should be built accordingly, with a fire in winter and a bed of tanner's bark in which the pot could be plunged. As befitted "a curious gardener" he advised how a gentleman could actually eat a banana. "It is a delicious fruit. First pare off the outward skin and then they emit a fine perfumed flavour and though they are not too terribly sweet at first, yet are of a most refreshing taste."

He added his verdict to Speechly's that it was the pineapple which had encouraged many of the nobility to build stoves and glass and that the pineapple was "now found in almost every curious garden."

Philip Miller, the custodian of the Chelsea Physic Garden, in his *Dictionary*, first published in 1731, lists as warm stove plants some 45 as follows: "Cashew, ahouia, alligator pear, allspice, arrowroot, banana, bastard cedar of Barbados, bastard locust of Barbados, bully-tree, button-wood of Barbados, cabbage-tree, cocoa-tree, cherry-tree of Barbados, calabash-tree, cassada, coconut-tree, winteranus, custard apple, date-tree, dumb cane, fiddle wood, fig-tree, flower-fence (*poinciana*) of Barbados, fustick-tree, ginger, guaicum (*Lignum vitae*), logwood, macaw-tree, mammea-tree, mancinel-tree, mimosa or sensitive-plant, nickar-tree or bonduc, palm-tree, papaw-tree, santa-maria, sour-sop, sugar-apple, sweet-sop, tamarind-tree, and tulip-flower or whitewood." In his glass cases which Miller said were built like stoves but allowed for more air than could be given in such structures, he grew ficoides, African sedums, cotyledons and other plants from the Cape of Good Hope, arctotis, osteospermum, royena and lotus. In his greenhouse for "plants too tender to live abroad in winter in England, but requiring no artificial heat and commonly called greenhouse plants," Miller lists 109 among which can be noted "begonia, calla (*arum*), campanula (bell-flower), rock-rose, convolvulus, heliotrope, hypericum, inula, ixia, physalis, ruscus (butcher's-broom), solanum and stapelia."

In a later edition of the *Dictionary* Miller adds to his list of stove plants, among others, amaryllis, asclepias, and buddleias, clover, cestrums, chrysophyllums, fleebanes, daturas, eryngiums, euonymus, mangosteen, lantana, opuntia, plumbago, rhus (sumac), robinia, saccharum (sugar-cane), sisyrinchium (earth-nut), smilax, spartium, vinca and urtica.

It can be seen from the preceding lists that the emphasis was still very much on economic plants, on trees and shrubs rather than on flowers, and that good and beautiful plants with which to fill the many new houses being built were difficult to come by. In 1730 we have Mr. Blackburn of Orford near Warrington, a pioneer botanist and stove-house man, writing to Dr. Richardson of Brierley, Yorkshire, actually saying just this. He wrote: "I have of late built some conservatories for keeping of greenhouse and stove plants which have in some measure answered my expectations; and I have applied to some of the gardens about London for plants of various kinds; for which they are so extravagant in their demands that I find I cannot fill my stoves and greenhouses without being at great expense."

He went on to beg for any plants Dr. Richardson could spare, and asked for information as to how he could get plants and from where. He was wanting geraniums and cistuses, he said, among other things.

Even when William Malcolm, who styled himself nurseryman and seedsman of "near Kennington Turnpike, Surrey," in 1771 issued his catalogue of *Hot House, Greenhouse plants, Fruit and Forest trees*, the emphasis had changed but little, and he listed sophora (Chinese pagoda-tree), stapelia, tamarindus (Indian date), tournefortia, turnera, volkameria, zamia (sago-tree), 26 different kinds of geranium, two cyclamen, a red and a white, *crassula coccinea*, portulaca, English fig marigold and 42 mesembryanthemums.

The Reverend Philip Hanbury, writing about the same time in 1770 of stove collections was so enthusiastic at the progress made that he claimed that in contemporary hot-houses and stoves could be found the different produce of the whole earth. "The wonders of the Almighty in the vegetative creation," he said, "may be here collected and His wisdom more and more displayed by the proper exhibiting of them."

He then put forward a plant list which is somewhat of an anti-climax, to say the least, for he spoke of the calabash, the coconut, dates, ginger, figs, mangoes, custard-trees, the bully-tree "and all sorts of palm-trees so much talked about by travellers."

A much more descriptive and pleasurable account of stoves and greenhouses in mid-century comes from the Reverend Joseph Ismay, Vicar of Mirfield, who visited Harewood House, Yorkshire, the home of the Lascelles family, in 1767. Writing for the benefit of his friends he told them:

"Among the fine collection of exotic plants and flowering shrubs in ye stoves and fire walls I observed the amaryllis or Jacobean lily, cistus or rock-rose, bananas in great plenty, various kinds of aloes, euphorbia, ficoides, egg-plants—the banana was ordered by Mr. Lascelles to be thrown out as too cumbersome and luxuriant for ye place—Cape jasmine (it bears a double white flower as big as a rose); palm-trees from the West Indies, it's a beautiful shrub; water melon, a species of passiflora from Barbados, a very curious plant; American viburnum, torch-thistle, opuntias, melianthus, or honey flower, Egyptian arum or colocasia, Ethiopian arum with a large white flower, this is a startling and beautiful plant; Arabian jasmine with a sweet-scented flower; fruit-bearing passiflora, it produced fruit last year in high perfection, the pulp is eaten with sugar and vinegar, it is more delicious than a melon and has much finer relish; Pancassian lily, a very stately plant; canacres (canna) or Indian shot; cereuscandens, the flowers are nine inches in diameter scarce continuing in full bloom twelve hours. The flowers of the creeping cereus are of a yellow colour; the East Indian solanum, Madagascar periwinkle, coix or Job's tear in flower.

"We saw ripe Alpine straws, kidney beans (ripe all winter), grapes nearly ripe, cucumbers, oranges, figs and peaches as large as crabs."

At Kew under royal patronage, where in 1761 Sir William Chambers had designed and built the Great Stove, now no more, but the largest in the country at the time with its 114-foot run and 60-foot bark stove running down the centre, the contents were not at all exciting by Victorian standards.

The plants recorded by the Kew authorities were the date-palm, *Ficus elasticus*, coffee arabica, *Piper negrum*, true aloes of South Africa, agave species (American aloes), sanseviera species, dracaena species and a few cactaceae.

At Upton in Essex Dr. John Fothergill, as we have seen, had a stove and greenhouse collection well ahead of its time, but even at that it was not particularly colourful, although in the language of the day it would be "curious" enough for all tastes. There can be little doubt that most of the plants established in the doctor's collection were grown from seed and that any listed as trees or shrubs would be in the small seedling stage for a considerable time.

While there is little point in listing every plant the botanical doctor grew, there is a *Hortus Uptonensis* assembled on his death in 1780 which can be consulted and this gives an excellent idea of the scope of the collection. It should be noted that many of the items listed as stove or greenhouse plants are today grown as hardy or half-hardy herbaceous border subjects.

Apart from the usual economic plants, aloes and succulents the flowers listed included *Achillea aegyptiaca*, alliums, alstromerias, amaryllis (16 varieties), *Anemone thalictroides*, five anthericums, three antirrhinums, arums (seven), asclepias, *Begonia obliqua*, bignonia (including *indica*), *Buddleia globosa*, one calceolaria, *Camellia japonica*, three campanulas (*aurea*, *rotundifolia* and *americana*) and Cannas represented by *indica*, *variegata*, *lutea* and *glauca*.

The list continues with ceanothus (*africanus* and *asiaticus*), two chrysanthemums (*flosculosum* and *maritimum*), four cinerarias and thirteen cistus or rock-roses. There were ten varieties of convolvulus and eight of the crassula family, while the crinums (lily asphodel) were represented by seven. Instead of the present wealth of varieties there were but two cyclamen (*indica* and *odoratum*), followed in list order by the daphne (*indica* and *cneorum*), twelve heaths, *Erigeron foetidum* and erythrina (*herbacea*, *corallodendron* and *picta*).

The fritillarias (*regia* and *nana*) were stove-house then as were two gardenias; genista (*spinosa*, *canariensis* and *candicans*) were greenhouse too.

Thirty-two geraniums made a brave show but there was but one gesneria (*tomentosa*). Two other bright subjects were seven gladiolus and one *Gloriosa superba*, while the seven varieties of hibiscus would certainly add to the colour of the stove collection.

Adding scent as well as colour were *Heliotropium peruvianum*, *Halleria lucida* (honeysuckle) and three jasmines (*grandiflorum*, *azoricum* and *odoratissimum*). Following alphabetically were three hypericums and five iberis with Ixia, a brave show with 15 varieties; *Ixora coccinea*, six lantanas, two lavendulas, one lavatera, two lobelias (*longiflora* and *coronopifolia*, both listed as stove subjects), two lychnis (*coccinea* and *coronata*) and as for the fig marigold (the mesembryanthemums), the doctor possessed what must have been the finest collection in the country with varieties totalling 68, and his collection of mimosa was an excellent one too with 26 varieties noted.

The Star of Bethlehem (*ornithogalum*) and the African chrysanthemum (*osteospermum*) followed with the oxalis (wood sorrel) growing in the stove. The pancratium or Mediterranean lily which Dr. Fothergill called narcissi, numbered seven and it is interesting to note represented Carolina, Ceylon, Mexico, Gibraltar, Africa and the Carribean. The Passion flower had a good showing with nine different varieties and one other in this alphabetical category was physalis (winter cherry) which was at that time grown in the stove. Plumbago (*zeylanica* and *scandens*) were side by side with plumeria and then there were two ranunculus (*alpestris* and *aconitifolius*) and four rhododendrons (*maximum*, *ferrugitifolius*, *ponticum* and *hirsutum*). There was but one rose in the collection, *Rosa indica*, as well as one rudbeckia (*laciniata*). Salvia in nine varieties are listed with scabiosa four, and scillas with two. Two sedums (*cepaea* and *rubens*) came under the greenhouse heading as did four sempervivums. The senecio collection totalled six, while the solanum, both stove and greenhouse varieties, numbered twelve.

The blue throatwort (*Trachelium caeruleum*) and tournefortia, tropaeolum, trollius, the globe flower here represented by *asiaticus*; two verbenas (*bonariensis* and *indica*), vinca (*rosea* and *alba*), Xeranthemum (*retortum* and *fulgidum*) and xylophylla (*longifolia* and *latifolia*—love flowers) rounded off the flowers.

The almost single-minded devotion to economic fruit and plants—for Dr. Fothergill's collection with his few flowers was most unusually comprehensive—lasted right into the first years of the next century and in 1810 we have Sir Joseph Banks still thinking primarily of fruit-forcing when he came to write so enthusiastically on contemporary greenhouse practice, although he was critical of the cult among the nobility for earlier and earlier fruit-forcing. He complained that stoves and greenhouses had hitherto been too frequently misapplied under the name of forcing-houses to the vain and ostentatious purpose of hurrying fruit to maturity at a season of the year when the sun had not the power of endowing them with their natural flavour.

He thought at the time of writing that there was a better feeling, a wiser regard for the true nature of the hot-houses and greenhouses coming along, however, and went on to say: "We [now] have peach houses built for the purpose of presenting that excellent fruit to the sun when his genial influence is most active. We have others for the purpose of ripening grapes, in which they are secured from the chilling effects of uncertain autumns; and we have brought them to as high degree of perfection here as either Spain, France or Italy can boast of. We have pine-houses also in which that delicate plant is raised in a better style than is generally practiced in its native intertropical countries; except, perhaps, in the well managed gardens of rich individuals who may, if due care and attention is used by their gardeners, have pines as good, but cannot have them better than those we know how to grow in England."

Sir Joseph then waxed eloquent, indeed prophetic, to forecast bigger, better greenhouses and stoves heated "by using naked tubes of metal filled with steam instead of smoke." He foretold "that ere long the aki and avocado pear and the litchi of China, the mango, the mangostan and the durion of the East Indies will be frequent at the tables of the opulent; and some of them, perhaps in less than half a century, be offered for sale on every market day at Covent Garden."

Although Sir Joseph was probably one of the first to realise what a revolution the introduction of tropic flora would mean to the British glasshouse scene, he was at this time, like so many more of his countrymen, placing the greater emphasis on fruit. His fellow-gardeners, not endowed with his imagination or knowledge, were still exclusively taken up with fruit and were unable to visualise the greenhouses and hot-houses of this country as repositories, as houses of defence, for plants and flowers from every part of the world.

A start had been made in this switch of emphasis from fruit to flowers, and the scene was changing almost imperceptibly, but changing it was.

For instance, Peter Collinson had received the first orchid from the Bahamas in 1731, a *Bletia verecunda*. In 1798 the number of exotic orchids at Kew had risen to 15, by 1813 it was 84 but by 1850 over 1,000. Indeed the introduction of new flowering plants was a most exciting feature of Kew, as it was of the country as a whole from about 1830 to the end of the century.

And what magnificent scope there was, what stupendous vistas in the very near future were opening up both for plant-hunters and greenhouse owners. As a contemporary writer put it: "At the beginning of the nineteenth century the floral treasures of great areas of the globe were still not only ungathered but unknown. All Africa, saving its Northern and Southern extremities, almost the whole of Asia, the two Americas with the exception of the Eastern seaboard of the North— all these remained practically virgin fields, open to the plant collector."

And it was excitingly true at this time that one packet of seeds received by Kew from its collectors abroad carried with it the possibility of some floral wonder which untravelled European eyes had never seen before.

In 1805 when Alexander McDonald published *A Complete Dictionary of Practical Gardening* his list of stove plants contained a few flowers, among them being achyranthes, alstromeria, amaryllis, arum, basella (Malabar-nightshade), browallia, calceolaria, erythrina (coral-tree), ferraria (black iris), *gloriosa superba*, haemanthus (blood-flower), heliconia (false plantain), hydrangea, polianthes (tuberose) and verbena.

But *Punch* about the same time, in a delightful little piece of doggerel characterising the achievements of the day and referring to one of the stove-houses at Oxford, still harped on fruits:

Look here is the banana abearing of its fruit,
And here you've got the plantain and the coco-nut to boot;
The coffee plant in berry you also here may see;
And likewise the prickly pear and the Ingy-rubber-tree,
The Ingy-rubber-tree.

Even in 1838 Paxton, writing of the greenhouse scene in winter, could still say of that period of the year: "Stoves and greenhouses generally are bare of interest and though here and there a solitary flower may develop itself, it seems rather to discover the general gloom than to enlarge the scene and delight the eye."

Loudon, that indefatigable Scot, sizing the position up statistically, estimated that at the opening of the nineteenth century 13,140 plants were in cultivation, of which 1,400 were natives. Of the rest he said 47 were exotics introduced during the reign of Henry VIII, 7 during the reign of Edward VI, and 533 during the reign of Elizabeth. Twenty came into the country while James I was on the throne, 95 during the Usurpation, 152 in the time of Charles II, 44 while James II reigned, and during the reigns of William and Mary 298, Queen Anne 230, George I 182, George II 1,770 and George III 6,756. During the first sixteen years of the century Loudon estimated that 156 plants came from foreign parts each year, yet this was a mere trickle compared with the deluge later.

But behind all the care and superb skill lavished on the growing of economic plants, on the growing and forcing of native and foreign fruits for the dessert, decoration and delectation of the tables of the rich, it must be recognised that a knowledgeable greenhouse tradition had grown up, invaluable experience had been gained, high skills learned and an all-round expertise secured which found British gardeners able and ready to tackle with confidence and quite amazing success the next step—floriculture in all its multifarious forms and its exquisite richness.

As the flow of horticultural treasures started, slowly at first until it reached flood dimensions, the glasshouse technique was there to handle almost any plant of any pretensions whatsoever from any part of the world. And that last phrase was literally true for as well as stove, tropical plant, hot, temperate, cool, aquatic and alpine houses, Loudon spoke of a "cold" house for the growing of musci, jungermanniae and other cryptogamous vegetation which grew in the lowest temperatures. His idea was to build a rustic vault of stone, let a stream of cold spring water run through its centre and arrange for water to drip from the roof continuously into the stream, or (and Loudon liked this idea better) build a pit with a double glazed roof facing north on which the coldest water was sprayed during summer from perforated piping. Canvas in two blinds was to be ready for excluding the sun and instead of flues lead piping was fed with the coldest of water from a lead cistern in which the water was kept constantly cold with ice to be renewed when it thawed. These pipes surrounded the house. Anyone, said Loudon, could then maintain a perpetual temperature of 32° F. which would allow the growing of lichen mosses and "all the most perfect plants which grow in the region of perpetual snows."

It was not mentioned by Loudon, but what a place in those refrigeratorless days in which to keep the milk!

Yet despite this high-sounding talk from horticultural writers in the early years of the century of what could be grown—and there was no lack of self-confidence among either these gentlemen or the gardeners who read them—the floral beauties of the world had still to be seen under glass in British hot-houses, stoves and greenhouses.

The Gardener's Magazine of 1833 was still giving high praise to one Mr. M'Gilligan, the purser on board the East India Company's ship the *Orwell*, who alone had managed to bring a few Chinese azaleas alive from Canton to London for Mr. Knight, although his fellow-officers on board who had formed a combine to do the same had met with dismal failure. Of nine variegated types two had reached these shores alive, of a score of double red ones six had arrived safely, yet previously a Mr. Reeve had shipped 500 azaleas and never got one to England alive.

Mr. M'Gilligan was certainly a horticulturist's treasure and it is to be hoped he was rewarded suitably, although of that there could be little doubt for the *Magazine* of the same year—1833—mentioned that at White Knights the Marquess of Blandford had planted rare foreign shrubs and trees which had cost him anything from ten to twenty guineas each, and went on to say that the noble Marquess's bill with Mr. Lee of the Hammersmith Nurseries in 1804 actually exceeded £15,000.

Yes, there was money for anyone who could bring an exotic alive and vigorous from the other side of the world to this country. All the magnificent glasshouses up and down the country and all the accumulated gardening skills of generations were waiting, although they could never have guessed it, for Dr. Nathaniel Bagshaw Ward, a London amateur naturalist and doctor practising amidst the dirt and smoke of dockland in the East End of London.

In 1824 Dr. Ward buried the chrysalis of a sphinx butterfly in some moist mould in a wide-mouthed bottle covered with a lid. "In watching the bottle from day to day," wrote the doctor, "I observed the moisture which during the heat of the day rose from the mould condensed on the surface of the glass and returned whence it came; thus keeping the earth always in the same degree of humidity. About a week prior to the final change of the insect a seedling fern and grass made their appearance on the surface of the mould."

Those particular plants, which revolutionised the carriage of exotics from all parts of the world, were to live four years in the bottle before the lid rusted and rain water got in during the doctor's holiday.

Thinking of the phenomenon, Dr. Ward put the whole thing down to "a moist atmosphere free from soot or other extraneous particles; light, heat, moisture and periods of rest and change of air."

In June 1833 Dr. Ward filled two cases modelled on his bottle, in other words a miniature greenhouse suitably strengthened for travelling, with ferns and grasses and sent them to Sydney, Australia. The cases landed with the plants in perfect condition, but the return journey was the real test, and in February, 1835, the case was filled with plants including *Gleichenia microphylla* and *Callicoma serratafolia*. The temperature conditions through which this Wardian case travelled with its precious cargo were those of 90° to 100° in the shade at Sydney, down to 20° at Cape Horn, an inch of snow at Rio with a rise of 100° to 120° on crossing the line. Eight months later in the Bristol Channel with a temperature of 40° the plants were found to be healthy and vigorous and for the first time ever a Gleichenia had been brought back alive.

The Duke of Devonshire was an early pioneer in the use of Wardian cases and received by them an *Amherstia nobilis* from India in perfect condition. With these simple but most effective plant-lifesavers in the hands of collectors the way was now wide open; the houses, the skills, the money, the demand were there and the flood-tide of exotic flora started to flow free and untramelled to Great Britain from the rest of the world.

Loddiges, the great London nurserymen, specialising in exotics, told Dr. Ward in 1842 that since 1835 they had used more than 500 of his cases for transporting plants from all parts of the world, and whereas previously they had always lost up to 190 out of 200 on long overland treks and on sea voyages, they now averaged 19 out of 20 alive and vigorous.

Dr. W. J. Hooker of Kew was kind enough to let the doctor know that his cases "had been the means in the last 15 years of introducing more new and valuable plants to our gardens than were imported during the whole preceding century."

There was another factor too which played its part in the definite change-over at this time from botanical and economic collections to a horticultural wonderland of riches. There was a definite and widespread move against the cult of fruit-forcing, of succulent and botanical collections, a sense of frustration and boredom with the old régime which was voiced by many horticultural writers of the time. This inspired and encouraged the enthusiastic owners of greenhouses to burn, bury, or dispose in any way of their old collections and to secure the new and dazzling wonders from abroad, whatever the cost.

Against the once prevalent taste for succulents, a gardening author of the time wrote: "Succulents cannot in general be considered beautiful; they are curious and some of them oddities and of forms which surprise at first sight; but who takes that pleasure in contemplating the leafless stapelia or the grotesque cactus (however extraordinary the flowers of some species may be) which he does in looking on a rose or camellia?—none whose tastes are not vitiated or singular, or who do not look solely with the eye of sense. One or two curious or ugly objects, however, may be admissible to show that there are such things," he conceded.

When Joseph Sabine, one of the first secretaries of the Horticultural Society, drew the particular attention of both collectors and member growers to the importance of seeking out and of cultivating tropical fruits, Mr. Loudon went right against the stream to say: "It [fruit-growing] seems to deserve the attention of retired persons of solitary habits, the aged or inactive. By presenting an end to be attained it may serve as a gentle stimulant to those such as from indolence or bilious complaint are apt to sink into a sort of torpid unenjoyed existence." Those who did prefer a greenhouse full of fruit-trees in pots, he said, had tastes which could not be considered either elegant or refined.

He had his say, too, as early as 1824, about the unimpressive, nay boring, appearance of so many of the greenhouses he saw up and down the country. "If there is such a thing as fine foliage, showy flowers, brilliant colours and elegant shapes, then three-fourths of the plants which require to be grown in greenhouses have no claims to these appellations. On the contrary we affirm that three-fourths are plants of meagre foliage, obscure, dingy flowers and uncouth straggling shapes."

That was fighting talk for his day and Mr. Loudon went on to suggest that such plants should be left to the botanists. Indeed,

M

he went further, to suggest that nurserymen in recommending plants acted as if all gardeners were botanists and again got in a broadside against greenhouse-owners and their displays with this tirade: "This is the reason why we see so very few greenhouses that present a gay assemblage of luxuriant verdure and blossoms; on the contrary they are generally filled with sickly, naked plants in peat soil, with hard names, which one half of the people of taste and fashion, and nine-tenths of mankind in general care nothing about." His suggestions for greenhouse displays were collections of heath, geraniums, camellias, salvia, polygalo, diosma, daphne, statice, fuchsia, aloysia, acacia, melaleuca, nerum, trachelium, lavendula and podalyria.

M'Intosh too, while congratulating owners for maintaining such excellent establishments and for their tastes in exotic botany, a taste which, he said, kept busy such a tremendous number of collectors in all parts of the world, was harsh upon the old tastes a-dying. "The mania," he said, "for accumulating species instead of forming judicious selections of good flowering plants has produced very baneful effects in the English garden, not only by excluding old and good plants, merely because they had long been denizens among us, but by introducing many which had no other merit to recommend them than novelty; how many of the plants of New Holland cultivated are scarcely worth the pot in which they grow?"

Dr. Ward and his cases changed all that, collectors were despatched post-haste to virgin country to seek out and send home any possible floral beauty or curiosity they came across. Competition among collectors and their firms reached fever heat and neither money nor human endeavour was spared in the race to transfer the flora of the known world to the greenhouses of this temperate island.

These introductions with their differing needs of heat, moisture and artificial conditions generally, brought precision to glasshouse nomenclature so that in men's minds an accurate definition could at last be given to orangery, stove, hot-house conservatory and greenhouse.

The orangery took over its original definition as a heavy masonry shelter for the preservation of trees and shrubs of the citrus family, while the greenhouse was properly defined as a structure specially devoted to the cultivation of plants that did not require a very high temperature in which the contents were in pots ranged on stagings for display effect. The house was not attached to any other building, unless it were a back wall, though more often than not a greenhouse was defined as span-roofed, well ventilated and with heat to go no higher than 60°.

The conservatory, it was said, was the highest in grade of all plant structures and required the greatest care and nicety in keeping. In it plants were grown in beds and the beds had always to show colour and plant form throughout the year. Ideally, and as the word was now understood, all conservatories should be attached to homes as a place of pleasant resort.

The stove or hot-house was a lower-built house requiring a considerable degree of artificial heat for plants from the tropics in the region of palms, tree-ferns, bamboo, orchids, aroids and ferns. Plants, depending upon the size of the house, were grown either in beds or on staging and in suspension from either bark or pumice.

13

THE EARLY PLANT-HUNTERS

IF the eighteenth century was the century of the botanist then the nineteenth was that of the dilettante, the century of the wealthy amateur who wanted a show not a lecture. The herbaceous, the alpine and the hardy aboraceous were not the fashion of the nineteenth century, nor were the arid plant families collected for dissection, nomenclature and cataloguing. The new introductions had to come in a blaze of colour, in a form bizarre, in a structure curious and in a novelty unique; nothing more or less would do.

All this then added up to one fact, and one fact only: hitherto untravelled parts of the world would have to be combed for plants and those plants to attract the new enthusiasts would have to be tender exotics needing not only the protection of glass but artificial heat as well.

One would like to record in definite terms that the Victorian glasshouse tradition and expertise triggered off the new and far-flung searches for horticultural novelty, and while to some extent with firms like that of Veitch it did, historical records show that many factors gave rise to the New-Look Glasshouse era.

Firstly there was bound to be, as in so many affairs of man, a swing of the pendulum, and in the gardening world this was from scientific botany to showy floriculture; secondly such travels as Captain Cook's, Humboldt's, Sir Joseph Banks's, and, later, Banks and Cook together on the latter's spectacular and awe-inspiring voyage of circumnavigation, had opened magic case-ments to a wonderland of colour, to a sub-tropic and tropic flora very much to be desired; thirdly the Victorians desired to match the known floricultural wealth of their foreign Empire, now that it could be accomplished, by miniature examples of that wealth at home and preferably as part of that home. They wished to fuse, and did, the metropolitan with the tropicopolitan.

With the introduction of the Wardian case not a single barrier stood in their way. Exploration, missionary travel and imperialist expansion reached a new high peak throughout the Victorian era and from the very nature of things these movements were largely confined to the tropical and sub-tropical rather than to the temperate regions of the world, and, equally naturally at first, to the lower altitudes of these regions. Forrest, Ward and Wilson and their collections of high, hardy Himalayan beauty were still to come and their introductions were many years ahead.

So it was that a large proportion of the foreign plants introduced to this country in the first three-quarters of the nineteenth-century were tender exotics all needing a great deal of care and protection only to be offered by the heated covered garden. And, with the new and efficient techniques of building, heating, cultivation and transport at a height of perfection never before experienced, with hundreds of acres of new British glass ready for the filling, with vast wealth at hand to do this, with a fashion which created an insatiable demand for horticultural newness and novelty, is there any wonder that the nineteenth century was the golden age of botanical discovery?

In the nineteenth century it has been well said that the English garden was furnished with the wealth and power of Empire as it had never been furnished before. Horticultural knowledge and botanical science became the right and enjoyment of all, a great industry was created; while beauty and diversity were brought to the gardens of the ordinary people, to the gardens of the wealthy it brought a surpassing loveliness. Never before or since has this country seen such an influx, such a golden flood of floral beauty as during the nineteenth century.

How the professional gardeners ever kept pace and up to date with it all must be a major mystery of horticultural cultivation, for the great majority of these clever growers and highly skilled plantsmen that they were, had never so much as put a foot outside this country—one could almost add, outside their conservatories and plant-houses.

They lived for their plants and could never in their wildest dreams have imagined the strange dim-lit world of a tropical jungle, the intense light and airiness of an Andean mountain peak, the steamy, sapping heat of tropical swamps or the lush green world of an Amazonian tributary.

Yet they grew the new marvels, they succeeded with the hitherto unheard of horticultural curios and ethereal beauties of an alien floral world. And they did this because the collectors, the plant-hunters, those intrepid, hardy travellers of the world, not only collected, but noted, observed, drank in the life background of the plants they tore from their tropic beds and passed on their knowledge lucidly, with a gardener's loving skill and care, to their colleagues in British stoves and greenhouses.

Once in the care of these clever hands all manner of devices and all the gardener's guile and art were directed to making both new seeds and plants feel completely at home. To assist germination, seeds were soaked in hot water, oxalic, malic or muriatic acids, covered in the pulp of rotten fruit or put into milk; they were laboriously chipped, heated, cooled, and stratified, while plants which had made incredibly long journeys through many varied climates were cosseted and blanketed, watched with a maternal care, and given every comfort for their successful growing in a strange land and in strange surroundings.

Johnson in his history had railed against the lack of skill with plants of the gardeners of Charles's, James's, and William's reigns, "nothing near commensurate to the pains which were taken to collect them." Why, he asked, did they not read about the new plants coming in so that they would have some idea of the climate they required and of the heat and cold they would suffer without harm?

He could never have said that about the nineteenth-century plantsmen. There must have been some dismal failures, as for instance with the early orchids and nepenthes brought over before a technique of cultivation had been developed, but by and large they pitted their professional reputation and skill with over-whelming success against critical odds to rear the new gems that had but recently seen the light of day in an English conservatory and potting shed.

But who were the second string of this highly successful horticultural partnership, who were the heroes and martyrs who risked health, limb and life for what one might, in some contexts, call a whim or a fad—that so prevalent and insistent urge for a man of wealth to be able after his evening meal and later, at the cigar-smoking stage, to enjoy by a peaceful stroll in the midst of beautiful, colourful, curious and rare tropic plants in his own conservatories and hot-houses?

This was not the context, however, in which early plant-hunting can rightly be put. For among men of vision like Banks at Kew, Sabine of the Horticultural Society and the principals of the commercial nursery firms who were clever enough to be just ahead of their time—Lee and Kennedy, Loddiges, Barr and Stroud, and later Veitch—there was a determination that the long-held floral beauty of foreign climes should be wrested from those climes and made to flourish here.

Many knew intimately the great herbariums of tropic and exotic flora which had been amassed over two centuries of assiduous and painstaking botanic travel. They knew the rich floral veins were there, that a floricultural wonderland lay just around the corner, and that skill, determination and sheer pluck would see that wonderland on their own doorsteps. Thus, at first the motivation was scientific, in other words men of the calibre of Banks wished not only to know a plant botanically from the dry-as-dust specimens in the libraries but to handle it, as it were, in the flesh and to cultivate it. But soon these highly colourful new creations of nature seen for the first time became a popular attraction and everyone who could wanted to give a home to these glamorous foreigners of the natural world.

Probably no other age than the Victorian, brimming over with scientific and philosophical ideas from the writings of such intellectual giants of natural history as Darwin, Hooker, Huxley and Lindley (who charged the already vital intellectual air with a new vigour), would have made the superhuman effort to collect and introduce, as it did in a matter of some eighty years, almost every worthwhile greenhouse subject of foliage and flower from every conceivable part of the world.

As for the collectors themselves, at first they were very much servants of their masters, trained in the Kew school or under a Paxton, but such a career as it developed later must have been a magnet for any man of spirit and imagination who could see, as we can see now when we look back in retrospect, those vast new floral worlds to conquer, a chance to explore unheard-of jungles, climb to peaks never reached before, to travel to the odd corners of the world; in other words a challenge to any man of action who loved and sought adventure and the high romance of travel in virgin territory on the unblazed trails of the world of green nature in the raw.

But what a toll of human life this tremendous task took! "To collect and send home the riches of these tropical regions," said Mr. Williams at the time, "is a work of much cost, and is attended with great difficulties and danger, in the prosecution of which many highly intelligent and talented travellers have fallen victims either to the pestilential climate, the wild beasts of the country, or the treachery of, in many instances, the equally wild aborigines."

For the sake of the orchid-fancier alone the record lists as victims to the Victorian orchidaceous craze no less than eight collectors who perished—Falkenberg in Panama, Klabock in Mexico, Wallis in Ecuador, Schroeder in Sierra Leone, Arnold on the Orinoco, Degance in Brazil and Brown in Madagascar.

For the nineteenth-century plant-hunter the world was definitely his oyster. Plant-hunters in some degree there had always been and even the Egyptians in their tomb portraits and bas-reliefs show armies and generals returning from foreign affrays bearing among their spoils not only prisoners, precious jewels and livestock but plants and trees. From the land of Punt they brought Queen Hatshepsut thirty-one frankincense-trees to plant at Thebes. The 3,450-years-old sculpture shows the tree roots carefully packed in reed baskets while the trees as a whole are slung on two carrying poles between a pair of slaves. Alexander is recorded as always having among his spoils of war plants and trees he fancied from the foreign lands he conquered.

But these were mere incidents in what, thousands of years later, became a global scramble for Nature's riches. In a minute degree compared with later centuries there had been even in medieval times some traffic between Europe and Britain, confined though it was to herbs and other medicinal plants and exchanged between the monasteries and the monks in charge of the gardens there.

The Crusaders also brought back a few Mediterranean species to be planted in the miniature gardens of castle keeps, so that some degree of continuity was maintained. In Stuart times the Tradescants carried on the tradition. John, the elder, gardener to Queen Henrietta Maria, travelled in Europe during 1620 in search of curious greens for my Lord Salisbury, and later, in 1637, travelled to Russia, botanising in the Archangel area before going off ostensibly to fight the Corsairs but actually to see what plants he might find and what seeds he could bring home from Algerian shores. He was successful with at least one introduction —the Algerian apricot.

But the growth generally of such plant commerce was slow, as was to be expected when one considers all the hardship and difficulties of travel in those early days.

A colourful figure in this early portrait gallery of plant lovers and travellers was Captain William Dampier, who combined piracy, navigation and botanising in a most chequered career. We catch a glimpse of him in August, 1699, with Pepys and John Evelyn at dinner being asked to look out for new plants on his travels. And travel he did, being the first Englishman to reach Australia. It was on some off-shore islands there that he found and recorded the plant later named after him, a firm Victorian conservatory favourite, *Clianthus dampieri*.

But the positive inspiration and spur to the art came from that small but highly influential coterie of Royal Society characters in the eighteenth century, the botanical and natural philosophy scholars whose urgent demands sparked off the plant-hunting adventure; to be exemplified later by the Sherards, Sloane, Bute, Petre, Collinson, Uvedale and Fothergill.

William Sherard, it may be remembered, when he was Consul in Smyrna, sent many plants and seeds to his brother James, who was famed for his garden of exotics at Eltham in Kent. Once again, these beginnings were modest enough, for traffic was almost solely between influential glasshouse owners in this country and missionaries, friends and relations abroad as well as with captains of sailing ships on the high seas. It was well said in those eighteenth-century days that no missionary or traveller ever dreamed of returning from foreign travel empty-handed of plants or seeds; it would almost have been to deny the privileges they had been granted to travel.

There is a record showing Lord Bute busy about the Prince of Wales's horticultural affairs and wanting to find worthy inhabitants for His Majesty's new stove at Kew, writing in 1750 to correspondents in Asia, Africa, America and Europe hoping they would find time to send home seeds as well as despatches. It was definitely an age of show the flag and filch the seeds. Lord Derby writing from Knowsley in 1738 to Leyden was desirous of having a camphor-tree.

A story of James Lee, of the Hammersmith Nurseries, gives an illuminating picture of the times. Strolling through Wapping in 1788, he saw in the window of a poor house a plant he wanted to secure, by hook or by crook. It was *Fuchsia coccinea*. He found it belonged to a sailor's wife, being a souvenir of one of his voyages to foreign parts, but she did not want to part with her treasure. Lee was determined, however, and bid up to £80 before the woman weakened and let him have the plant on the understanding that she could have the first two rooted cuttings, which Lee said he would propagate. He was successful, not only with the two rooted cuttings but with thousands more, and subsequently made a handsome profit on his initial outlay.

Probably that dear, honest, upright Quaker, John Bartram of Philadelphia, should have pride of place as the first serious systematic collector of plants for gain. From 1735 onwards he roamed the virgin American forest lands, plains and mountains for plants and seeds for Peter Collinson, the London wool merchant of Mill Hill, who supplied plants and natural curiosities for the collections, amongst others, of Lord Petre and the Dukes of Richmond and Norfolk. It is true most of his finds were hardy species, but many of his seed collections were germinated and first saw the light of day in English stoves and hot-houses. His first earnings as a plant-hunter was five guineas per box of plants and seeds to reach Collinson.

He made epic journeys into the American wilderness, deep into hostile Indian country, often with no companion but a horse on which he carried everything, two panniers for plants and seeds, his blanket roll and a few luxuries of civilisation such as flour; for the rest he had to live off the country. On one trip to Maryland he covered eleven hundred miles in five weeks, and what rough, uncharted miles they were—up mountains, down river gorges, across lakes. As we can see by the hundreds of letters which passed between them, he was urged on by the pressing demands of Collinson and his patrons for new plant discoveries, the new wonders of the New World.

For forty years Bartram collected seeds and plants and corresponded botanically with Collinson, and most of the time Collinson was able to secure patrons for his plant-hunting protégé at ten guineas a year—not much, one might say, for the tremendous effort put into the task, for quite often there might be but one haul per year of plant discoveries, and communications were of the most protracted, letters between the two taking as long as five months to exchange. As to the precious cargoes lashed to the deck of ships in casks and boxes or swinging precariously in tarred sacking in the hold, in danger from rats and rough handling by sailors and from weather, salt spray and extremes of climate, disappointments were many. On one occasion Collinson clambering aboard a sailing ship found a rat had made a nest and reared a brood inside one of his precious plant boxes; they were there, pinkly stirring, when he opened it in London River after it had been packed at Bartram's forest home, amidst a ruin of roots and dead greenery.

Lord Macartney's embassy to the Chinese Emperor in 1792 was accompanied by two botanists and as a result a mere trickle, an appetising indication of the beauties yet to come from this hitherto closed book of a people long horticulturally minded, began to find its way to England.

As early as the beginning of the century, James Cunningham, a surgeon for the East India Company, started a tradition of plant-collecting and botanising in Asiatic territory and, like the many other floral martyrs to follow, experienced some of the dangers and difficulties which dogged the horticultural adventurer in foreign countries. In Macassar his station was attacked and all the Englishmen on it but Cunningham murdered. He escaped with his life only to spend two years as a prisoner. When next heard of he was in Bhutan but fell foul of the natives there and was expelled from the country. He embarked for a home he never reached, for he died on the voyage. Among his credits are *Hibiscus manihot, Sapium sebiferum* and *Rhus semialata,* the latter being raised from seed at the Chelsea Physic Garden.

In any history of Chinese plant-collectors the many nameless Jesuit missionaries must always rank for praise, for they regularly sent home live plants and seeds to Europe, to Paris particularly, from the nursery gardens in the neighbourhood of the Imperial court at Canton. Of those who were known, Pierre d'Incarvaille might stand as an example of this long and honourable line, for his seed collected from the vicinity of Pekin ultimately reached Philip Miller at Chelsea after it had travelled either via Paris or by trade caravan across Siberia to St. Petersburg and from there on to London.

With Sir Joseph Banks came the era of worthwhile, full-scale efficient collecting; indeed Sir Joseph could well be called the father of modern plant-hunters as well as the instigator of modern techniques of plant-hunting.

For fifty years his word was law at Kew and it was there that, from his own personal experience, he trained his young men for the arduous tasks they had before them, when they would leave the confining walls of this London garden to travel thousands of unknown miles into virgin territory to seek the natural treasures he knew existed there. His own plant-hunting experiences were particularly varied and wide, no one's more so, for certainly in his day there could not have been anyone who had seen more of the natural curiosities and wonders of the world. His first journey had been to Newfoundland and found him in difficulties when his ship barely escaped shipwreck and much of his collection was thrown overboard. What he did bring back were a few hardy shrubs, but he was soon to leave the bleakness of the north for scorching equatorial heat and lush floras.

In 1768, with Dr. Solander, he joined Captain Cook for four years in the *Endeavour* (368 tons) on his great voyage of exploration to the Pacific. He had many adventures, including near-death from exposure—two of his companions did die—one January night in 1770 on Tierra del Fuego. Visiting New Zealand and the east coast of Australia, Botany Bay was named to commemorate the new and unusual forms of plant life they had already seen. In the Malay Archipelago both Banks and Solander had malaria and nearly lost their lives, and again came near shipwreck on a coral reef off the Queensland coast.

It was this epoch-making voyage recorded in the journals of both Banks and Captain Cook, that for many of their countrymen were first opened up vistas of a flora both unexpected and desirable. It was also his experiences of plant discovery and of the hazards and dangers of plant-preserving and plant introduction which gave Sir Joseph a pre-eminent position as the tutor of tyro-hunters in the Kew academy.

How about the early pupils? Curiously many were Scots or of Scottish descent, and a remarkable fact about these young men who were to travel the seven seas, explore unknown lands and overcome hardships, dangers and a thousand and one vicissitudes completely unknown to modern travellers, all for a few handfuls of green wealth, was that they were mostly drawn from the humble, uninspiring background of cottage gardens. Their paternal acres, or their masters', bound them physically and mentally, and yet, raw from a slow-moving, conventional countryside, they were launched to the ends of the earth.

The first and one of the most successful pupils of the Kew school was Francis Masson, an under-gardener and a native of Aberdeen, whom Sir Joseph had trained in plant- and seed-preservation. He was sent out in 1772 at the age of 31 to South Africa and on 9 April he set out, as had his master before him, with Captain Cook. From the Cape Masson made many journeys into the hinterland, trekking by ox wagon. On one occasion, when accompanied by Carl Thunberg, he got as far as the Sundays River before being turned back by hostile natives. He was near to being murdered in the Table Mountain area where he just escaped capture by a gang of runaway slaves and had to spend the night in a rough hut with but a clasp knife as his sole weapon against the convicts and the many wild animals he knew were prowling nearby.

Masson had two spells in South Africa, but between 1778 and 1782 he was off to new pastures with a trip to Madeira, the Canaries, the Azores and the West Indies, seeking plants and seeds for Kew. This trip was full of dangers and adventures for the times were troubled and there was war with the French, so that Masson found himself fighting at the fall of New Grenada when he was captured by the enemy and naturally lost his collections.

Between 1783 and 1785 this intrepid Scot was diligently finding new treasures for his master in Spain and Portugal. Those being the days of sail, when any journey depended upon the vagaries of climate and wind, Masson on his next expedition took four months to get to North America, some part of the delay being due to his capture by French privateers.

Eventually he reached his destination, but communication between Canada and Kew was difficult and nothing is known of his work there, the next record being that of Masson's death at Montreal due to an extremely cold winter in 1805.

Masson's was a rich harvest which within a very few years would grace the Victorian conservatories—fifty Cape heaths, proteas, ixias, *Strelitzia regina* named for his Queen, and the gay cinerarias, the latter all due to his introduction of the wild senecio, among some other four hundred assorted discoveries.

A floral martyr who surely must have suffered more for plant introduction than any of the long and heroic line was David Nelson, another Kew trainee, sent with Captain Cook's third expedition on the *Resolution* when he collected plants from some of the most inaccessible parts of the world during the four years of the voyage. He saw sub-antarctic islands, the tropical islands of the South Seas and the desolate shores of the Aleutian Islands and the Bering Sea. There is every reason to think he acted as a supernumerary ship's carpenter for Cook, and he certainly built his own plant-cases. From Tasmania, New Zealand and Australia Nelson had the distinction of taking home for the first time specimens of the acacias and the eucalyptus, one of them becoming a great conservatory favourite, *Acacia verticillata*. It was on this voyage that Captain Cook was murdered by the natives at Tahiti and the senior botanist, Dr. Anderson, died leaving Nelson in charge, but his real tribulations were still to come.

In 1787 when Captain William Bligh was appointed to command the *Bounty* and to take breadfruit from Tahiti to the West Indies and some to be brought home to Kew, Nelson was put in charge of the plants with William Brown, another Kew gardener as assistant. The long, tiring journey was uneventful; it took a month for the ship to struggle round Cape Horn, and Tahiti was reached and breadfruit loaded along with seven hundred other plants collected for Kew. Then commenced that eventful journey to the West Indies. In April the crew mutinied, Captain Bligh was dragged from his bed along with Nelson and sixteen others, thrown into an open boat and left in the midst of the wide Pacific. Everyone must know the story of how Bligh steered for Timor, 4,500 miles away, and reached it forty-seven days later. On the way there the little party rested for several days among some of the coral islands of the Australian coast and trouble sprang up again, to be settled only by Bligh challenging the leader to a duel. During this trouble the Captain recorded in his log that only David Nelson assisted him. On the day after this gallant boat's crew reached Coupang on 20 July, 1789, Nelson died of "an inflammatory fever" brought on by the privations and hardships he had undergone.

When Nelson was so roughly thrown overboard, his assistant William Brown found himself with sword in hand forced to take the side of the mutineers. Subsequently he sided with Fletcher Christian and found himself on lonely Pitcairn. He proved himself a most useful member of the community with his knowledge of botany and natural history, but a dispute arose about ownership of some land in the colony and poor William Brown was another Kew martyr, for he was shot whilst tending his own garden.

George Caley, son of a Yorkshire horse dealer, was another colourful Kew collector who travelled thousands of miles in Australia, often alone, to help fill the glasshouses of Sir Joseph. Barely twenty-one when he got tired of his ten shillings a week as an under-gardener for Banks, he was given a free passage to Australia as a plant-collector. The next ten years he spent in exploring the lonely valleys, forests and mountains of south-western Australia sending home epacris, grevilleas and acacias.

He is a part of the early Australian records, not only because many of the most delightful of that Continent's wild flowers are named after him, but because his rugged character could not be contained in the conventional regulations of the Colony; he fell out with another now famous Yorkshireman and pioneer, Samuel Marsden, the missionary who introduced sheep-farming to Australia, and got himself into the public eye by his amorous attentions to convict women. His work, though not spectacular, was highly praised by later collectors. He came home in 1810, bringing with him Dan, his aboriginal manservant, a great natural curiosity in those days.

Later he accepted the post of Superintendent of the St. Vincent Botanical Gardens. He fell out with the local gentry when he closed the gardens—a fashionable parade ground—on Sundays and he complained that the paraders were helping themselves to plants and seeds.

14

PLANT-HUNTERS EVERYWHERE

By the opening of the nineteenth century the tradition had become established and collectors were part and parcel of the horticultural and glasshouse story. In 1810 Allan Cunningham, a Scot by descent, came from his Lincoln's Inn chambers where he was reading law, to help William Aiton, the curator at Kew, with the second edition of his *Hortus Kewensis.*

Within four years he was off to South America with James Bowie, sailing for Rio in October, 1814, and, in those days of long journeying, making landfall in late December. After a spell collecting in the vicinity they set out across country and reached São Paulo after a month of rough-going. This was all more or less virgin territory to Europeans and the pair had the honour of being the first collectors to explore the Organ Mountains area in southern Brazil. They were successful in finding and introducing various species of the curious bromeliaceous plants, including *Billbergia pyramidalis*, the true *Cattleya vera* and *Verbena erinoides*. After some two years in South America, Cunningham was ordered to New South Wales and spent four months on the high seas making the journey in the convict ship *Surrey*, not, one would have thought, an auspicious start or an appropriate conveyance for one of His Majesty's plant-collectors.

Once in New Holland Cunningham set off on incredibly long journeys into the interior—he actually circumnavigated the island continent twice—and for fourteen years in those early colonial days served Kew and his royal master well. He crossed the Blue Mountains and on one trip saw Caley's Repulse, a pile of stones set up to show where George Caley, many years before, had had to turn back on one of his early exploratory trips, almost starving and with no hope of more provisions. Cunningham, however, pushed ahead, far beyond the Blue Mountains and finally, after

N

suffering from thirst and many privations, reached Bathurst, a journey of twelve hundred weary miles in nineteen weeks, "under most trying conditions," as an early biographer so mildly puts it. Exploring Norfolk Island he found *Capparis nobilis* and *Solanum bauerianum* as well as *Lagunaria patersonii*, a warm greenhouse subject. Another trip to the north-west was made on the *Mermaid* of 85 tons—the imagination boggles at journeys of such magnitude in what could have been little more than a bath-tub with sails. As it was, provisions ran short and the ship had to make for Timor before reaching Port Jackson safely. He sought plants on this trip in the Illawarra region and in Tasmania.

After seventeen years he came home to Strand-on-the-Green, but wanderlust fighting against the boredom of one ordinary day following another won at last and in 1837 he returned to Sydney as Colonial Botanist and Superintendent of the Botanical Gardens there. He soon relinquished the post, however, when he found that the greater part of his staff consisted of forty convicts and that his main task was to market-garden for the colonists with the convicts as under-gardeners. He went to New Zealand again but failing health compelled his return to Sydney and on the eve of returning to England he died in June, 1839.

His younger brother Richard was one of the early martyrs who died a violent death for the sake of green treasure. He, like his brother, started as an assistant at Kew on the clerical side, working on the *Hortus*, but in May, 1832, took over at Sydney the Botanical Garden to which Allan Cunningham went later. Eighteen weary weeks at sea under sail in the confined cabins and ships of those days with the extremely limited provisioning which accompanied those conditions preceded his entry into Australia. Shortly after arrival he was off again in H.M.S. *Buffalo* to New Zealand and in March, 1834, returned to the Bay of Islands. Early the next year he was travelling again with a party who were out to find the source of the Darling River. Cunningham, who had a facility for losing himself completely when intent on finding new plants, became parted from the main party on 17 April. For four days his companions sought him and found a track showing he had gone off leading his horse. Next the body of the horse was found and on 2 May Cunningham's handkerchief.

Later, much later, the tragic story was pieced together, when it seemed that some time after 24 April, exhausted by hunger and thirst, Cunningham had fallen in with a party of friendly natives who had fed him. It was thought that during the night he had become delirious and the natives, becoming frightened, had murdered him on the spot.

At this time China was a magnet for plant-collectors. It had been a sort of unattainable Shangri-La for centuries and many attempts had been made to get plants out of the country, but with little success. Kew first tried with a man on the spot in 1803 when Sir Joseph Banks sent William Kerr who stayed in Canton, as he was compelled, visiting the nursery gardens in the city and nearby and sending home mostly hardy plants until he sailed for Colombo in 1812.

Where Lord Macartney's mission of 1792 had failed, the Government of the day hoped that in 1815 conditions would be different and the Chinese authorities more helpful to the botanist, and so sent out Lord Amherst, with every hope of success. Again Kew supplied a man, Hooper, to look after a plant cabin for the preservation of living specimens—a pre-Wardian Banksian case. Very little came of the venture, however, and three hundred lots of seed were lost when Hooper's ship was wrecked on a coral reef in the Banka Straits.

One of the most successful introducers of new plants from China came under the category, mentioned previously, of convenient friends abroad. He was John Reeves, chief tea inspector for the East India Company at Canton. From 1812 until 1831 his services to British exotic horticulture were invaluable. Not only did he find and send plants and seed to this country but he experimented with ways of preservation, devising a type of Wardian case and a method of potting-up plants from the interior and establishing them at the port before despatching them on their long sea journeys. He himself spoke to the captains and crews of East Indiamen and taught them how to care for their precious green cargoes. Although the records of his introductions are not numerous several authorities feel that many species linked with the names of various ships' captains and travellers from this country should actually be credited to Reeves.

As it was, his was the task of establishing a tradition of skill in handling and conserving plants which he passed on fully by letter to influential horticulturists in this country, and verbally to many official collectors who were sent out there. John Parks, of the Horticultural Society, recorded that he had much for which to thank Reeves.

By 1820, however, on the death of Sir Joseph Banks, the most commendable self-imposed task which Kew had undertaken from late in the eighteenth century, of searching the foreign strands for exotic booty and beauty came to a halt and the initiative was taken over then by the newly-formed Horticultural Society, later to become the R.H.S.

Probably because of the colourful pictures of China conjured up by the introductions of Reeves and his associates, the Horticultural Society chose that country for their first attentions, and in 1821 sent out one of their gardeners, John Potts, who made a round trip to China on an Indiaman. His trip was not particularly spectacular for he lost forty chrysanthemum varieties on the voyage home, but he did manage to introduce several varieties which became later conservatory favourites, some camellia varieties, *Ardisia punctata*, and seed of a delightful greenhouse subject, *Primula sinensis*, as well as other greenhouse plants.

An almost identical trip a year later was made for the H.S. by John Damper Parks. Neither he nor Potts could be strictly called plant-hunters for Mr. Reeves, Dr. Livingstone and Mr. Beale, Englishmen resident in Canton, put chrysanthemums and other Canton nursery plants in their way and all the two collectors had to do was to see them safely home, not, of course, an easy task before the invention of the Wardian case. Parks did however manage to get home safely with thirty varieties of chrysanthemums, *Camellia reticulata* and four other varieties of that much-loved Victorian buttonhole flower, as well as the orchidaceous

Coelogyne fimbriata and the first *aspidistra* to be seen in England. This plant was *A. punctata*, however, not the popular Elatior to be seen on every house window-sill after it was introduced later from Japan in 1835.

The chrysanthemum, sent in many varieties from China at this period, was, of course, a face-saver for the early Victorian conservatory owner, for here was a comparatively easy plant which would fill with colour the drear dull days of October, November and December.

The first of the cultivated species introduced was a purplish crimson flower called *Chrysanthemum morifolium* which went to France from China in 1789. Later it came to London, along with a "sport" Changeable white, with mingled white and crimson florets.

Between 1798 and 1808 seven new varieties were introduced direct to England by Sir Abraham Hume of Wormleybury, while the nurserymen James Colvill, of Kings Road, Chelsea, and Lee and Kennedy, of Hammersmith, introduced new varieties due probably to Reeves or friends stationed either at Canton or Macao. Then Barr and Brookes had Joseph Poole, who had been attached to Lord Amherst's embassy, collect chrysanthemum varieties for them in Canton, and such was the influx of this plant that by 1826 Joseph Sabine, then secretary of the Horticultural Society, was able in a lecture to describe no fewer than sixty-eight new varieties.

Because he kept a very full and informative diary and journal of his travels, probably the best known of the early nineteenth-century plant-collectors is David Douglas, who collected in North America for the Horticultural Society. He too was a Scot, the son of a Scone stonemason, who started life as an apprentice gardener before going into private service with Sir Robert Preston at Valleyfield on the Forth. Then this keen young man at twenty-three years of age found himself chosen by the Horticultural Society of London to follow in the footsteps of John Bartram. Unlike John, who at least had the advantage of starting out on his travels from home territory, Douglas's second trip to the North Pacific coast of America took eight months' sailing, beating round Cape Horn and tacking up the length of the South and North American coast before sailing up the Columbia River to land at Fort Vancouver.

Although David Douglas's collections were mainly of the hardy type of flowers, trees and shrubs and not greenhouse subjects, the man's adventures are worth recalling to give some indication of the very real hardships of a plant-collector's life in the nineteenth century.

Quite soon on his second trip he found himself with one companion at Cape Foulweather, battling out a torrential storm under a few pine branches and a soaking wet blanket, without food. For several days, suffering from exposure, they managed to exist on the roots of *Sagittana saggitifolia* and *Lupinus littoralis*. The plants they collected on this trip had to be thrown away because with a lack of food they had had to dismiss their Indian porters and eat some of the berries they had collected for seed.

Douglas had his share too of extremely dangerous and uncomfortable canoe journeys with overnight stops at most inhospitable camps; one he records was such a steep slope that it was impossible to pitch a tent. They could not sleep anyhow as four hundred Indians surrounding the camp were, as Douglas wrote, "inclined to be troublesome." By morning light he fought a battle of wills with an Indian whose aimed bow and arrow failed against Douglas's own nerve and quickness with a gun.

This was not his last exciting encounter with Indians by any means and on several occasions only his dour Scot's courage and native wit kept his scalp from a chief's collection.

In the September of 1829 Douglas, who had covered some 10,000 miles on foot, in canoes, and on horesback, set out for home with eyesight failing from attacks of snow-blindness and generally a sick man, owing to the many hardships of travel.

In October of 1829, however, he was off again to the Pacific Coast of America and once again suffered disaster when his canoe was dashed to pieces on the Frazer River losing him all his specimens, all his equipment, and almost his life.

In 1883 he left for the Hawaian Islands where, judging by the contents of his journals, he was enjoying the sunshine, the climate generally and the change of scene. But his holiday was soon over for while he was gently botanising he fell into a pit dug by the natives to catch wild animals, and on 12 July, 1834, his mangled body was found alongside a wild bull which had gored him to death.

Next, Robert Fortune, an imperturbable Scot, was taken from his superintendency of the hot-houses at Chiswick to collect in China. His books and journals show him to have been an almost heroic figure, risking his life on many occasions for plant introductions. He was responsible for the importation of many greenhouse favourites, including *phalaenopsis* and other orchids, chrysanthemums, azaleas, that favourite conservatory forcer—*dicentra spectabilis* (the bleeding-heart) and *Jasminum nudiflorum*, at first always grown under glass.

Travelling to China on the very heels of the Treaty of Nanking, when the Chinese were still sore from defeat, Fortune sailed to this territory new to the botanist on 26 February, 1843. He had already had a tussle with the Horticultural Society Committee, who wished him to travel in what he considered dangerous territory with only a life-preserver as his means of defence, but he argued his way to a fowling-piece and pistols. He was asked in a long memorandum by the Society to look for nepenthes, yellow-flowered camellias, the mandarin orange, varieties of celosia and amaranthus as well as a host of hardy plants, trees and shrubs.

Fortune was lucky in his collecting in that, under the Treaty, Europeans for the first time had more of China in which to travel and were able to visit Hong Kong, Amoy, Fuchow, Ningpu and Shanghai, and an area of some forty or fifty miles around those towns. Fortune's first voyage (and he made four to the Far East) took four months out of Liverpool to reach Victoria, Hong Kong. And there he was "as if I were a being from another world" to thousands of the Chinese, who followed him everywhere, peeped into his pockets or examined his clothes or just gazed wonderingly at him. Quite often the crowds got ugly and shouted at him "*Wyloe*" (be off), but he used to put a bold face on the matter, stare them out, walk in among them and usually things were settled amicably.

In Canton on one occasion he wandered off seeking plants with neb-cap and umbrella, a picture of Victorian sporting elegance, unknowingly wandered into the roughest quarter, where he was set upon by thieves who took his cap, umbrella and money. A few score of the Chinese surrounded him and trapped him in a walled cemetery with one door, which they shut, but he bowled over the nearest man and, like a human battering-ram, threw himself through the door. He fisted his way through another hostile crowd which had gathered outside, was half stunned by a brick which hit him at the back of the head, and he escaped amid a hail of stones.

Another time the merest chance rescued him from falling into a pit similar to the one in which Douglas lost his life. On yet another long journey, like so many of his compatriots in an alien climate, he fell ill from a fever and for days was insensible at intervals on his rough bed in the junk in which he was travelling on the Min. "It seemed hard for me," he wrote, " to die in a land of strangers without friend or countryman to close my eyes or follow me to my last resting place."

But that was not to be, for Robert Fortune had need to forget his fever and fight for his life against even greater odds. His junk was attacked by eight pirate junks; but let Fortune tell his own modest, unvarnished tale:

"I got out of my bed ill and feverish as I was, and carefully examined my firearms, clearing the nipples of my gun and pistols and putting on fresh caps. I also rammed down a ball upon the top of each charge of shot in my gun, put a pistol in each side pocket and patiently waited results. By the aid of a small pocket telescope I could see, as the nearest junk approached, that her deck was crowded with men; I then had no longer any doubts regarding her intentions. . . I knew perfectly well that if we were taken by the pirates I had not the slighest chance of escape, for the first thing they would do would be to knock me on the head and throw me

overboard as they would deem it dangerous to themselves were I
to get away. . . . I was surrounded by several of the crew, who
might well be called 'Job's comforters,' some suggesting one
thing, and some another; and many proposed that we should
bring the junk around and run back to the Min.

"The nearest pirate was now within 200 or 300 yards of us, and
putting her helm down, gave us a broadside from her guns. All
was now dismay and consternation on board our junk, and every
man ran below except two who were at the helm. I expected
every minute that these also would leave their post; and then we
should have been an easy prey to the pirates. 'My gun is nearer
you than those of the Jan-dous' said I to the two men; 'and if
you move from the helm, depend upon it I will shoot you.' The
poor fellows looked very uncomfortable, but I suppose thought
they had better stand the fire of the pirates than mine, and kept
their post. Large boards, heaps of old clothes, mats and things of
that sort which were at hand, were thrown up to protect us from
the shot; and as we had every stitch of sail set, and a fair wind, we
were going through the water at the rate of seven or eight miles
an hour.

"The shot from the pirates fell considerably short of us, and
I was enabled to form an opinion of the range and power of their
guns, which was of some use to me. Assistance from our cowardly
crew was quite out of the question, for there was not a man
among them brave enough to use the stones they had brought on
deck; and which perhaps might have been of some little use when
the pirates came nearer. The fair wind and all the pieces of sail
which we had crowded on the junk proved of no use; for our
pursuers, who had much faster sailing vessels were gaining
rapidly upon us. The shot this time fell just under our stern. I still
remained quiet, as I had determined not to fire a single shot until
I was quite certain my gun would take effect. The third shot which
followed this came whizzing over our heads and through the sails,
without, however, wounding either the men at the helm or
myself. The pirates now seemed quite sure of their prize, and
came down upon us, hooting and yelling like demons, at the same

time loading their guns and evidently determined not to spare their shot. This," he remarks nonchalantly, "was a moment of intense interest. The plan which I had formed from the first was now about to be put to the proof; and if the pirates were the cowards which I believed them to be, nothing could prevent them from falling into our hands. Their fearful yells seem to be ringing in my ears even now, after this lapse of time, and when I am on the other side of the globe.

"The nearest junk was now within thirty yards of ours; their guns were now loaded, and I knew that the next discharge would completely rake our decks. 'Now,' said I to our helmsman, 'keep your eyes fixed on me, and the moment you see me fall flat on the deck you just do the same, or you will be shot.' I knew the pirate who was now on our stern could not bring his guns to bear upon us without putting his helm down and bringing his gangway at right angles with our stern as his guns were fired from the gangway. I therefore kept a sharp eye on the helmsman and the moment I saw him putting his helm down, I ordered our steersmen to fall flat on their faces behind some wood, and at the same moment did so myself. We had scarcely done so when bang, bang went their guns, and the shot came whizzing close over us splintering the wood about us in all directions.

"Fortunately none of us was struck. 'Now, mandarin, now, they are quite close enough,' cried out my companions, who did not wish to have another broadside like the last. I being of the same opinion, raised myself above the high stern of our junk, and while the pirates were not more than twenty yards from us, hooting and yelling, I raked their decks fore and aft with shot and ball from my double-barrelled gun. Had a thunder bolt fallen amongst them they could not have been more surprised; doubtless many were wounded and probably some killed. At all events, the whole of the crew, not fewer than forty or fifty men, who, a moment before, crowded the deck, disappeared in a marvellous manner, sheltering themselves behind the bulwarks or lying flat upon their faces."

This was the end of number one pirate but another came on,

shooting. Fortune played the same trick and from thirty yards gave them the contents of his double-barrelled gun again; exit number two!

However, not three days later, and still in his bunk with fever, Fortune was told by his captain that pirates were on their tail again. This time he showed resource as well as courage and dressed as many of the native crew as possible in his European clothes and bade them hold belaying-pins like guns. But a broadside from the pirates foiled this scheme, so in true British fashion he withheld his fire until he could see the whites of the pirates' eyes and let 'em have it fore and aft as usual and with the usual effect.

For the transit of his collections Fortune had the advantage of the Wardian case and actually took three cases of living plants out with him to report on their efficiency.

On coming home from China in 1845 he divided his collections; eight cases of living plants he sent home by one ship, and duplicates of these plants and others he took home himself in eighteen glazed cases on the poop of the *John Cooper*, where he cared for them from 22 December out of Canton to tying-up in London River on the 16th of May next. All the plants, he reported, arrived in excellent health.

Fortune also gave a detailed description of the method of transmitting plants by "Ward's case." Firstly he gave general hints; well-seasoned wood, tight glazing, nine inches of soil topped with moss for most plants, bark on which orchids had grown to be nailed to the wooden back of a Ward's case; feet six inches high; always put the plants and soil into the cases about a fortnight before the sea journey to get them established; stowage on poop-deck always, although main or mizzen top could be used unless the Captain objected, which he usually did. As to his own treatment of the eighteen valuable cases and their precious hoard on the 16,000-mile journey from China to London Fortune wrote how each day he withdrew an end slide of each case, after decks had been washed, to remove dead or damping leaves and to dress and clean the soil surface. Off Madagascar he made all the cases tight with putty and never opened them again until round the

Cape. At St. Helena he gave them all the fresh water they could take; other collectors travelling with their cargoes took their cases ashore here for a breather, fresh air, fresh water, and to steady their legs! At the equator he opened a sash for the plants to receive refreshing showers and for the sun to ripen the wood. Steering for the Western Islands with the vessel close hauled and spray dashing over the decks the cases were closed again and he sailed into the English Channel and home with 215 of the 250 plants he put into the cases alive and vigorous.

Later Fortune hunted in Japan when that country was opened to Europeans after being closed to them for two hundred years. Here once again he was to travel in a hostile country as something of a curiosity to the natives, and a nuisance to officialdom. He was pursued by crowds of amazed Japanese wherever he went; armed guards had to escort his visits to nurseries to keep him from robbers, murderers and a hostile populace generally. Nursery-men were frightened of him and sent scouts out to watch for his appearance, when they would scurry back to bolt and bar the gates. But he travelled widely despite the dangers and difficulties, being so determined to make his way to the gardens of the interior that on one occasion knowing not one word of the language he travelled as an itinerant native beggar.

He was able to tell his countrymen, now exceedingly well versed in the art of conserving plant life under glass, that they had little to learn from the Japanese in this respect. Despite that country's traditional love of flowers for centuries, their efforts at the conservation of exotics and tender plants were crude indeed. At Yedo the nursery gardens, he reported, had sheds and rooms with thick mud walls ranged with shelves for the storage of pots, delightfully called winter-houses. The thick mud walls were on the north, east and west sides, the sheds had mud roofs, and to the south were open with a light framework of wood on to which was pasted paper as a glazing. In exceptionally cold weather straw mats were thrown over the paper, or sometimes, Fortune re-corded, hot-air chambers and furnaces furnished artificial heat in a crude way.

As from China, Fortune sent his plants from Japan in Wardian cases and he writes, in a touching aside, of his treatment of "some special favourites which I did not like to trust to the long sea journey round the Cape, which were brought home by the overland route under my own care. One of these is a charming little saxifrage, having its green leaves beautifully mottled and tinted with various colours of white, pink and rose. This will be invaluable for growing in hanging baskets in greenhouses or for window gardening. I need not tell you how I managed my little favourites on the voyage home; how I guarded them from stormy seas and took them on shore for fresh air at Hong Kong, Ceylon and Suez; how I brought them through the land of Egypt and onwards to Southampton. More than a few fellow passengers by that mail will remember my movements with these two little hand greenhouses."

A pioneer collector in new territory was John Gibson, the "intelligent gardener" Paxton and the Duke of Devonshire sent out to India in 1835 to bring back the *Amherstia nobilis* and as many orchids as he could find. Gibson had the advantage of the Wardian case as well as the enthusiasm and wealth of Chatsworth behind him and he was equal to both. With Dr. Wallich of the Calcutta Botanical Gardens as a most valued ally Gibson brought home a cargo which according to the doctor was "a truly princely one and I trust commensurate with the liberality to which it is entirely due." There were twelve cases to stand on the poop-deck at £2 each and four to accompany Gibson in his cabin. Gibson, who was given an excellent name by Wallich, travelled in Bengal and Assam and was "dreadfully miserable in consequence of the dreadful heat of the climate. The thermometer is seldom lower than 80 degrees but often as high as 98 or 99 degrees and with this not the least breath of air."

Later, in the good ship *Zenobia* off Plymouth in July, 1837, he was able to write with joy and pride of a mission well accomplished. To the Duke he wrote:

"The plants are in the most beautiful order and preservation. The collection of plants is very extensive occupying 13 glazed cases and several open boxes, several of the plants contained in the latter were procured at the Cape of Good Hope. There are also a quantity of orchidae attached to branches of trees which have been suspended in my cabin during the whole voyage from India. I can enumerate upwards of 100 species of orchidae independent of the other fine plants which comprise the collection and which were not in England when I took my departure from India. . . ."

Even with this magnificent collection at Plymouth Gibson's troubles were not over; he was worried lest delay and rough handling would undo all his work before he got his unusual and rare cargo to Chatsworth, so he suggested that a fly-boat should be standing ready on a nearby canal basin to take everything in their cases by water to Cromford in Derbyshire, whence they could be forwarded to Chatsworth the same day as their arrival. This scheme was carried out with evident success.

A later expedition mounted from Chatsworth, when Paxton sent two young gardeners, Wallace and Banks, to California, ended in tragedy. They were both drowned in the Columbia River before they had chance to make any plant-collections.

15

FLORAL GOLD

BEFORE considering in detail the work of the professional plant-collectors, who followed the pioneers of the art, the men sent out by the great nursery firms which either expanded or were founded to meet the urgent and unprecedented demand for tender exotics suitable for greenhouse, conservatory and stove, it behoves the present-day glasshouse-owner to spare a thought for these intrepid horticultural adventurers who sought plants before the days of air travel, canned and concentrated foods, lightweight equipment, distant rail-heads or penetrating motor-roads, accurate maps, the horseless carriage itself, preventive medicines and drugs, scientifically designed clothing and business-sponsored conveniences of travel generally.

It is obvious from the start that only the toughest and most resourceful of men had to be chosen, for once they left these shores they were entirely on their own and were more often than not required to live by their own resources like any Robinson Crusoe.

Your man needed a strong character and had to be quick-thinking, resourceful, persistent, mother-witted, with an unusual blend of the scientific approach and eminent practicality. He would have to face such risks as cannibals, wild animals, bubonic plague, typhoid and malaria, and would have to undertake almost superhuman journeys over difficult terrain.

In some parts of the South Americas he had "a jaguar every yard," as one collector wrote, or hostile natives who wished for nothing better than a white man's head, complete with all the usual Victorian hirsute appendages, to shrink as a number one trophy. Travellers found themselves in jungles where "the appearance was strange in the extreme—large trees covered with creepers and parasites and the ground encumbered with fallen and rotten trunks, branches, leaves, the whole illuminated by a glowing vertical sun and reeking with moisture."

The silence of these primeval forests was fraught with apprehension, for at any moment it would be ear-splittingly torn by horrible screams as some beast or bird fell a prey to a marauding animal. In many forests the gloom under the great trees was unrelieved and snakes lurked underfoot. At one time there would be slimy mud up to the knees, at another there would be a difficult and dangerous scramble over rock and fallen trees.

Java and the Philippines, in orchid territory, and particularly the mountainous regions, was strange, unaccommodating country for the Kew or nursery garden apprentice. It was virgin jungle, needing natives with billhooks to preceed the collector to make a way through the impenetrable brushwood or tangle of leaves. Once in the jungle, wrote an early collector, "the tops of the trees form a mass so dense that no ray of sun ever shines through them. The ground on the sides of these mountains is always in a moist and slimy condition and is the habitation of millions of leaches. . . . At first when they fixed themselves upon me, which they did through my stockings, I set to work to pull them off, without regard to species, although warned by the natives of the impropriety of doing so. In a short time my legs were covered with blood and the wounds annoyed me with a kind of itching soreness for several days afterwards." In many tropical forest regions skin-boring ticks were prevalent, and their barbed proboscis had to be painfully extracted from the skin.

In India and near Tibetan terrain the searcher for plants had different vegetation and conditions to fight. There were altitude-sickness and temperature extremes to contend with, as well as the enervating effect which made wearisome the scrambles up steep paths through dense, deep-green dripping forests to the snow-line. Everywhere the soil and bushes, the travellers reported, swarmed with large and troublesome earthworms. In the evening after a wearisome day the noise of the great cicadas in the trees was deafening, and not at all conducive to sleep. There was always the danger, too, from charging rhino or wild boar; while deep in the jungle one might meet unexpectedly leopards, tigers and elephants. Wild banana and screw-pines spread their high barricades, huge cable-like lianas coiled round branches and hung perpendicularly down—a tiresome bar and hindrance to progress.

To ease communication, river journeys and side of river journeys were often made, but these were arduous and slow in the sort of gorge and torrent conditions the collector usually found on his travels. If a riverside path was used then, because of landslides or huge rock falls blocking the path, difficult scrambling detours up the steepest of gorge sides had to be made, quite often for many hundreds of yards. If a raft or canoe was used then porterages were often so frequent as to become almost ridiculous; one orchid-hunter reported on a South American trip that the canoe on the river-way demanded thirty-two porterages and as many loadings, for every time the canoe had to be man-handled over the rocks or round the rapids all equipment had to be lifted out and packages carried separately so that the heavy, long canoe could be manœuvred with safety on the steep, rock-strewn narrow paths around the obstacles in the stream.

Marching through unexplored country meant carrying food and equipment to meet most emergencies. Joseph Dalton Hooker, travelling in Sikkim and Nepal in mid-century, gave a remarkable if unusual picture for the time of a type of large-scale organisation. "My party," he wrote, "mustered 56 persons. These consisted of myself and one personal servant, a Portuguese half-caste, who undertook all offices and spared me the usual train of Hindu and Mahomedan servants. My tent and equipment, instruments, bed, box of clothes, books and papers required a man for each. Seven more carried my papers for drying plants and other scientific stores. The Nepalese guard had two coolies of their own. My interpreter, the coolie sirdar (a headsman) and my chief plant collector (a Lepcha) had a man each. Mr. Hodgson's bird and animal shooter, collector and stuffer with their ammunition and indispensables had four more. There were, besides, three Lepcha lads to climb trees and change the plant papers, and the party was completed by fourteen Bhutan coolies laden with food

o

consisting chiefly of rice with ghee, oil, capsicums, salt and flour. I carried myself a small barometer, a large knife and digger for plants, note book, telescope, compass and other instruments, whilst two or three Lepcha lads who accompany me as satellites carry a botanizing box, thermometers, sextant and artificial horizon, measuring tape, azimuth compass and stand, geological hammer, bottles and boxes for insects, sketch books etc. arranged in compartments of strong canvas bags."

Hooker had his fair share of adventure despite his flair for organisation, for on his second Himalayan trip he ran into trouble at the court of the Regent of Sikkim and was forcibly detained for twelve months during 1849, an incarceration which nearly ended in his assassination.

M. Oversluys, an orchid-hunter in Brazilian forests, told of areas where it was impossible to put down a shilling-piece between the red spiders and other insects. None of the five peons he took with him would climb for plants, as they were almost bitten to death by insects; two ultimately had mud-fever and another two ran back to their villages as they would not face the insects which they knew would attack them in hundreds when cutting orchids from their tree-top habitat. Some of the ants were nearly two inches in length and often travelled in wide swarms across the jungle floor.

The journeys with plants from where they were found to the coast ports for shipping home were difficult and protracted in the extreme, and a contemporary nineteenth-century collector in South America wrote to his firm in London to tell them of his personal difficulties.

"We pack them in boxes on mule-back here and then the convoy sets off through thick forest country for a ten-day march to Bogota. Here the cases are examined for damage and decay and repacked and put back on mule-back again for a six-days march to Honda on the Magdalena River where they are taken from the mules and packed on to rafts for a voyage of fourteen days to Sabanilla during which not a breath of wind stirs the air and the boxes on deck under blankets and palm leaves are soused with water all day. At Sabanilla there is probably a wait for days in sweltering heat for the Royal Mail. Then the boxes are stowed on ship board where their contents are spoiled with salt spray or, packed in the hold where they will ferment." No assurance company would insure such cargoes for which the collector and his assistant had probably risked life and limb. After all these tribulations there might be but one or two miserable pieces of plant left for the shipper when he unpacked at London docks. Sanders actually lost 267 cases on such a journey. In another ship Roezl despatched 27,000 *Masdevallia schlimii* representing a fortune; his company received two meagre bits, which they managed to sell for 140 guineas each.

Masdevallia davisii, which was found in Peru, 10,000 to 12,000 feet up in crevices of rock, was most difficult to transport, it was reported, for as they journeyed on mule-back for the two- or three-weeks journey from mountain-top to the seaport, each day's journey brought them into a progressively hotter climate. It was a rare plant too, growing in a most restricted area only, extending in territory but a few miles along the fringe of the mountains.

Of the great Victorian nurserymen who employed men like Roezl and expressly catered for the glasshouse-owner, Veitch was probably the most illustrious name in a formidable catalogue of famous firms and had, at any one time from 1840 onwards, the greatest number of collectors in all parts of the world. But other leading London and Home Counties firms were not lacking in initiative; Messrs. Mackies in the Kew Road and at Clapham had collectors in Australia; Mr. Tait of the Sloane Street nurseries, grew and sold many of the new plants sent from Mexico by Mr. Bullock; Conrad Loddiges and Sons, of Hackney, who were noted for the magnitude and extent of their hot-houses, were fortunate to receive the novel plants collected by Protestant missionaries from all quarters of the world; Shuttleworth and Cardew dealt in orchids from Colombia; Standish and Noble, of Bagshot, had Far Eastern plants in their houses. At any one time it was a peculiar pleasure for the plant connoisseur to be able to find a new and different plant in bloom at each establishment.

There was Hugh Low, of Clapham nurseries, who received the riches of the Borneo flora from his son Rajah Sir Hugh Low, who was out there on Government service in 1843; tropical plants, orchids and nepenthes came from this source. The business of Knight and Perry of Hammersmith was later taken over by John G. Veitch, the son of James. He travelled in Japan and the South Pacific to send home orchids, crotons and dracaenas.

To some extent, apart from the ostentation which was an integral part of Victorian class-consciousness rather than a vice to be decried, the many florid and larger tropic foliage plants, palms, tree-ferns and climbers, replacing the lowly ericas and geraniums, demanded the loftier, more spacious and more ornate glasshouses of the time—fitting crystal caskets for the floral treasure-trove of the world!

Something of that tropical wealth which flooded into the country during the last three-quarters of the nineteenth century can be gauged from the introductions of Veitch, whose firm was founded at Exeter in 1832. Of stove and greenhouse plants generally, they brought here 392 species including hoyas, acalyphas, aecthynanthus, alocasias, anthuriums, aristolachias, dieffenbachias, a total of 24 dracaenae, many calceolarias, *clianthus dampieri*, dipladenias, eranthemum, fuchsias, impatients, that beautiful hot-house climber *Lapageria rosea*, some new passifloras and *Plumbago rosea*.

Forty-eight pitcher-plants, the nepenthes, were established in British hot-houses along with no less than 117 exotic ferns and a grand total of 138 new orchid species.

As one glances back to the nineteenth-century scene it is most evident that the horticultural catalogue of today would have been a poor, uninspiring and colourless one but for the Victorian introductions. The Veitch's themselves wrote of their times and their introductions: "It would have been strange if results were not forthcoming when such practically virgin lands as certain parts of South America, Japan and central China were open to men of the calibre of the Lobb brothers in the early '40's and the late John Gould Veitch in the early '60's."

Actually it was as early as 1840 when James Veitch felt as a good plantsman and businessman that the demand for exotic plants was so great as to make it worth while to send out his own collectors. He despatched William Lobb, one of his gardeners, to Brazil. For four years Lobb, who first went to Rio de Janeiro, travelled thousands of miles in virgin country visiting Chile, crossing the great pampas lands of the Argentine, scaled the Chilean Andes and proceeded southwards to penetrate the great Araucaria forests.

His second trip to Brazil saw him prospecting in Valparaiso, southern Chile, Valdivia, Chiloe and northern Patagonia where he found and sent home *Lapageria rosea*, *Tropaeolum speciosum*, and many orchids as well as foliage plants. In 1849 he was collecting in North America and explored southern California, while in 1852 he collected on the Columbia River in Oregon. Next year he was searching the Sierra Nevada for the new and the novel and from the autumn of 1854 until 1856 he was collecting in California, where he died in San Francisco a year later.

His brother, Thomas Lobb, who collected for the firm from 1843 until 1860, was sent to entirely different territory. He was actually destined for Singapore, but when it was found that China was not hospitable to travellers at the time he was directed to Java and then to India, where he collected for three years. By the time he retired, less a leg which he lost as the result of exposure in the Philippine Islands, in some twenty years of plant-seeking he had travelled thousands of miles and visited on foraging expeditions the heavily monsoon-drenched Khasi Hills in Assam, north-eastern India to Lower Burma. He sent many fine orchids from the Malay Peninsula and North Borneo and was successful in introducing the first nepenthes to be grown in British hot-houses.

The present beautiful race of tuberous begonias, so much a feature of summer shows, are all due basically to Richard Pearce, another successful Veitch collector, who roamed the South

American continent between 1859 and 1866. He collected many stove and greenhouse plants in Chile, Peru and Bolivia and was the first to find and send home the tuberous begonia from which all the present magnificent modern specimens are derived. His plant was *B. boliviensis* collected in Bolivia in 1864 and of the other plants used for hybridising many were Veitch collected, some by Pearce himself; they were *B. pearcei, B. veitchii, B. rosaeflora, B. davisii, B. clarkei.* Pearce for instance sent *B. pearcei* from La Paz in 1865 and *B. veitchii* from near Cuzco in Peru at an elevation of 12,500 feet. *B. rosa* he found in the Andes in 1867 and *B. davisii*, also from Peru, was run to earth (or should it be sky) at 10,000 feet? The very altitude at which these plants were found indicates to anyone with imagination the arduous tasks the collector had to face to find his treasures.

Two other Veitch collectors met unhappy ends about this time, for David Bowman, who was sent out to search for stove-plants in Brazil in 1866, was set upon and robbed by natives just as he was about to sail for home. He lost all his possessions and died of dysentry in Bogota. Henry Hutton set out in the same year for Java and Malay, commissioned to find orchids, but two years later died as they delicately put it "of the climate."

Orchid wealth was now worth every effort to procure and Gottlieb Zahn tried his luck in 1869 for orchids and ferns in Central America but was drowned in Panama in September of that year. Fortunately for the glasshouse-growers and owners in this country, the supply of men was well-nigh inexhaustible so long as there were new and exciting orchids and stove-plants to be found, and hard on the heels of Zahn went George Dowton for orchids to Central and South America and the islands of Juan Fernandez.

In the same year that Dowton set out—1870—one of the most colourful and successful characters employed by Veitch, J. Henry Chesterton, who started his working life as a gentleman's gentleman—he was a valet with a difference, however, for he had travelled extensively with his master—was sent out to the Americas to bring back alive the rose-coloured odontoglossum, later named *Miltonia vexillaria*, and he did. Noted in the Veitch records as "an excellent collector of orchids," his keenness and enthusiasm for the job brought about his death. "Poor Chesterton's reckless spirit rendered him very efficient as a plant-collector," wrote his employers, "but he had been very ill yet his great spirit gave him high hopes of recovery and he set off up river to Puerto Berrio in Colombia, was taken desperately ill again and had barely been borne ashore by his native crew when he died."

During the time that Chesterton was seeking the orchid in Colombia, Gustave Wallis was seeking the flower in Brazil, New Grenada, the Philippines and Panama from 1872 for several years. He died of fever at Cuenca in Ecuador in 1878 but before paying his debt to the gods of orchids, he carried out some remarkable journeys and actually travelled from the mouth to the source of the Amazon, exploring that great river as well as some of the more important tributaries.

Another heroic figure who combined original exploration with plant-hunting was Walter Davis. Going out to South America in 1873 to collect in Peru and Bolivia he is recorded as having crossed the Cordilleras no less than twenty times, climbing up each time to heights of 14,000 to 17,000 feet. He is also remembered in orchid annals as well as geographical ones for having traversed the continent from one side to the other along the whole length of the Amazon valley.

Looking specifically for tropical flowering and foliage plants and covering a tremendous slice of ground in doing so, Guillermo Kalbrezer hunted on the west coast of Africa, in Colombia, Fernando Po, Old and New Calabar and the Cameroon Mountains and river. In 1876 he ran up against hostile native traders, great heat and frequent attacks of malaria, but managed to find that beautiful Anthurium—*veitchii*—with leaves three feet long.

So insistent were the demands for particular plants that from time to time Messrs. Veitch found it worth while to send out collectors to search for a single subject. For nepenthes, for years a difficult subject to transport and to acclimatise in this country,

F. W. Burbidge went to Borneo in 1877. The pitcher-plant had been discovered by Rajah Sir Hugh Low in 1851, but he had not been able to introduce it. Thomas Lobb had also tried and after an initial success the plant failed, but Burbidge succeeded. He was out in Borneo for one year.

New ground was broken by the firm, after James Veitch's early trip to China, when they sent Charles Maries to Japan and China. He collected in the Yangtze valley and travelled to Yokohama, Nikko, and north to Hakodate, then to Yezo and Sapporo in Hokkaido and on to Formosa. He suffered severely from sun-stroke. In 1879 he travelled up the Yangtze as far as the Ichang gorges, where he had trouble with the natives who robbed him of all he possessed so that he hurriedly made tracks back to Japan, but his journey resulted in his collecting seed of *Primula obconica* and in the introduction of *Hamamelis mollis*, said by one authority to be the best semi-hardy plant to come out of the Far East.

Another successful collector in his own line of nepenthes, stove-plants, ferns and orchids, was Charles Curtis who from 1878 to 1884 searched for Veitch in Madagascar, Borneo, Sumatra, Java and the Moluccas.

An unusual character employed by the firm in the East Indies, Burma and Colombia in the last few years of the century was David Burke, who on his expeditions lived more or less as a native, and by living under these primitive conditions brought on his early death; indeed it was only by a mere chance that his friends and employers heard he was dead. Some natives were able to tell a white traveller that he had died in the lonely squalor of a native hut miles upon miles away from a European settlement.

It was by the efforts of men like these, by the sacrifices, some-times of life itself, often of health, and always of the comforts of home and families, that nineteenth-century floral wealth was brought to these isles.

16

VICTORIA'S FLORIFEROUS REIGN

So it was that as Queen Victoria came to the throne the wealth and power of the Empire furnished the English garden as never before. It seemed almost as if the flora of the world signalled her accession and the opening of the Victorian era of British green-house gardening, the most magnificent and most floriferous epoch in the country's long and honourable history of horticulture.

The gardeners and nurserymen all strove, consciously or un-consciously, to match the era not only with hybrid nomenclature redolent of those opulent times but with plants to match the richness and graciousness of British country and aristocratic life. If ever there was a time when to have a jungle on the doorstep in stove-conservatories and aquatic-houses was completely in keeping with the general environment, then it was during Victoria's reign, when almost every younger son of the house-hold, and many more sons in those days of large families, were probably doing Foreign Office duty on far-off shores with a flora to match the one at home; when the very backcloth to living was florid and highly decorative in knick-knacked, antimacassared, wax-fruited, fussily-furnished mansions, with which the Asiatic forest flourishing outside the drawing-room door was so much in keeping.

If there had not been gardeners and greenhouses ready to hold those glittering horticultural prizes, then some of the abounding wealth and brassy splendour of the era would have found other channels into which to overspill but British and world horti-cultural history would have been the poorer for it. The tremen-dous progress made in plant-discovery and botanising could never have occurred, without the enthusiastic and wealthy private patron.

What a thrilling time it was, when even a prince of his

profession such as Paxton could record monthly new plants he had seen in bloom for the first time as he went his rounds of the great nurseries which had arisen to supply the ever-pressing need for newer and better plants for the thousands of stoves, conservatories and greenhouses which had been built. There was Rollison's of Tooting, Messrs. Chandlers of Vauxhall, Mr. Young of Epsom, Messrs. Henderson of Pine Apple Place, Mr. Knight of Chelsea, Messrs. Loddiges of Hackney, Mr. Low of Clapton, and, later, the firm of Veitch. At any of these great establishments there was always the exciting prospect of some utterly new experience, some completely new discovery of natural beauty never seen before by English eye. Time and time again Paxton could report in his journal: "This new and highly beautiful plant has recently bloomed."

Here for the people of temperate climes was displayed for the first time the wonders of tropical vegetation; the palms, aroids, lianas, orchids, were parts of vegetable nature completely outside the experience of all but very few tropical travellers, and as such their appearance in the tropical plant-houses, stove-conservatories or a palm-house of the time was one of the major wonders of the Victorian era.

Less than ten years after Paxton had complained of the lack of attraction of the conservatory in winter his master, the Duke of Devonshire, was writing enthusiastically to his sister, listing the wonders of the great stove at Chatsworth. This was in 1844 and among the inhabitants of the house he listed the orange, altingia (liquidambar) araucarias (the Norfolk Island pine), the Walton date palm, *Hibiscus splendens*, *Erythrina arborea*, palmettos (Sabal palms), plantains, the Walton *corypha umbraculifera* (the talipot palm), *Bougainvilaea spectabilis*, *Stephanotis floribunda*, *Sagus rumphii* (one of the sago palms), *Bambusa arundinacea*, *Sabal blackburniana* (the first stove-plant the Duke ever acquired "of Mrs. Beaumont of the North"), *Cocos plumosa*, the rose hibiscus and musae.

He then described "A sort of jungle" with cassias, *hibiscus mutabilis*, the banyan-tree, *Ficus elastica*, *Dracaena draco*, fourcroyas (Syn Furcraea), *Ficus repens*, and "beyond a hedge of aloes serving as fence to inquisitive public," the Duke recalled the climate of southern Europe. Here he had more oranges, brugmansias (*Datura arborea*), abutilons, mandarin oranges, *Chamaerops humilis*, Franciscea, Hopea, *Poinsettia pulcherrima*, *Cycas revoluta*, sugar canes, *Brownea grandiceps*, *Carica papaya*, and a zamia (sago palm).

This last, which came from the Tankerville collection, was an example showing how anxious were these new greenhouse fans to have their collections as complete as possible, for to secure the prize the Duke had had the Tankerville house which gave it shelter pulled down to get it out, a special carriage was invented to carry the arboreal monster over the roads to Derbyshire, and a turnpike was demolished at one place on the journey to allow the tree passage.

Then the visitor came to "the dismal pool," a Chatsworth nickname for the aquatic end of this giant glass case, where a limpid pool was surrounded by papyrus, arums, Chinese rush and *Hedychium coronarium*. Ranged around at "elbow height" were the South African collections, geraniums, succulents, and so on, which Baron Ludwig had torn from his garden at the Cape, "so charmed was he with the conception of the Great Conservatory," and sent to Chatsworth.

To set off the flowers the Duke and Paxton had also skilfully placed crystals; one described as "a mountain of light" was found during the making of the Simplon Pass, while others were from India and Siberia, and there was also an opalised stem of a tree brought from Van Diemen's Land.

If you can imagine looking across the great Palm House at Kew from its surrounding gallery for the first time, when every single leaf, fruit, stem, bark and flower below, around and above were utterly new, remarkable and almost unreal, then you may judge the effect of a Victorian stove-conservatory upon both its owners and its visitors.

Incredible sums of money went into these great houses to make the rockwork, the pools, the fountains, the walls, so artfully disguised with barks and pumice, look like a natural grotto or with skilfully planted undergrowth to mimic a jungle floor. So striking were these jungles under glass that in those open to the public visitors were hushed into silence and the atmosphere was almost one of bowler-hat-carrying, for the gentlemen at any rate.

Let us glance at some of the novel plants which were available to grace the crowded borders and fill the ornamental stages or climb round the spiral ironwork of the many-pillared pavilions.

There were the stately and colourful crotons (Codiaeum), of which some three or four species were first discovered in the East Indies in 1804. Here was a plant which matched the era: it graced the conservatory, gave colour and style to the groups, to the drawing-room and great-hall jardinières, and suggested an environment both affluent and aristocratic. The nurserymen saw to it that such a plant lived up to its early expectations and produced ever-new varieties. By 1884 Nicholson listed 69—with what high-sounding names! Queen Victoria, Disraeli, Baron Franck Seilliers, Crown Prince, Earl of Derby, Lady Dorothy Neville, Prince of Wales, Baron James de Rothschild, Princess of Wales, Princess Mathilda, Lord Balfour, Baron Alphonse de Rothschild and Baron Nathaniel de Rothschild! The patrons were not far to seek. And what a list of origins Nicholson gives, a graphic picture indeed of the terrain of the assiduous collectors: Polynesia, New Guinea, India, New Hebrides, South Sea Islands, Eromango.

Its colour-combinations were almost unlimited: light and deep yellows, orange, red, crimson, all shades of green, bronze, rosy crimson, ivory-white, pink, spotted, blotched, lined, variegated . . . and this after myrtles, coffee plants, succulents, and a eucalyptus!

Then there was the caladium, a plant symbolic of its time in its lush luxuriousness of colour and form. Introduced in 1773 with *C. bicolor splendens*, a native of Madeira, Nicholson later listed 133 different named varieties. Its massive heart-shaped leaves, its superb coloration was elegance embodied; it was a must in all conservatories.

The anthurium was another plant which came to the zenith of its glory during the Victorian stove and conservatory era, with some 150 species, representing the collectors' zeal in the West Indies, Costa Rica, Colombia, Ecuador, Brazil, tropical America, Venezuela, eastern Peru and South America.

There was the cult too for the novel and bizarre, exemplified by the insectivorous and pitcher-plants. Stove-houses were specially designed for the curious family of pitcher-plants, the nepenthes, from the swampy jungles of Singapore, Borneo, Sarawak, northern Australia, China, Madagascar and the East Indies. Nicholson lists 46 different ones, yet after its first discovery in 1789 nothing happened until 1874 when, from that time right up to 1884, new ones came flooding in as collectors searched their jungle haunts.

Think of the thrill of seeing *Nepenthes distillatoria* for the first time, towering twenty feet high in the Chatsworth stove, from the tangled two-foot-long leaves of which hung fifty pitchers, each six to nine inches long with a circumference of some five inches, or *N. rafflesiana* with green bottles densely spotted and speckled with bright red-crimson. The rim of the pitchers is reflexed and crimson broken, and in the pitchers there may be up to half a pint of water. *N. curtisii superba* has pitchers the ground of which is rich blood-red with longitudinal yellow-green streaks and marking, the rim dark red and the lid freckled with red on a yellow ground.

Here were exotics indeed!

Also for novelty there was the dionaea (Venus's Fly-trap) the leaves of which are sensitive and close upon insects as they settle on them; the saracenia, a miniature pitcher-plant to drown insects; or the pinguicula which traps the orchid-fly (*sicara pectoralis*) and the drosera from Australia and New Zealand whose leaves are covered with glandular hairs laden with dew to entrap insects making a landfall.

The *Begonia rex* from Assam reached these shores only in 1857 when Mr. Rollison of Tooting introduced it. Once again here was a plant to delight the connoisseur and one which matched in its rich and bizarre variety of leaf colour some of the magnificence of the Victorian background. It was another introduction with which the hybridist worked wonders and once again the nomenclature of its varieties spoke of its affluent patronage: His Majesty, Kaiserin Augusta Victoria, Princess Charles of Denmark, Our Queen, President Carnot, Princess Olga.

And what stove-house magic there was in the miconia. In *M. magnifica*, for instance, with its huge long, oval-shaped leaves tapering to a point and up to two feet six inches long and a foot across—the upper surface a lustrous green with a distinct ivory-like midrib criss-crossed with horizontal lines of intense white; the underside of the leaf, a rich purplish crimson.

Grown intensively by the Victorians also were the dracaenae, which, introduced as "Dumb Cane" (because of the bad effect it has upon the human vocal cords if any of the plant is eaten) many years before the Victorian era, sprang into new life during the great Queen's reign. From the beginning of the nineteenth century almost to the close from New Guinea, China, Mauritius, tropical West Africa, the Canary Isles, India, Java, Madagascar and the East Indies the new plants came in and again the nurserymen perpetuated the identity of their patrons in naming them: Prince Manouk Bey, Lord Wolseley, Victoria, Lord Roberts; the long lists read like a Victorian Who's Who.

Spoken of by some of the early writers of the stove era as the prince of stove-plants the ixora was also a favourite for the Victorian stove-house. Named, it was said, after a Malabar idol to whom the flowers of some of the species were offered in worship, it had just the right kind of romantic origins and from 1814 up to 1880 Nicholson records new species being sent from India, the Andaman Isles, the East Indies, Java, New Guinea, the South Sea Islands, China, Bengal, Ceylon, tropical Asia, Africa, the Moluccas and Tenasserim.

It was among the most handsome and gorgeous of stove-plants, evergreen with showy clusters of flowers, white, yellow, orange, rose and scarlet.

For the ardent collectors there was, as well as the flamboyant, the meticulously beautiful—the rare miniature jewel, as against the large florid decoratives, and this was found in the anoetochilus, a stove evergreen from Java, India and Ceylon. The leaves were small, some two to three inches in length, but as a Victorian writer said: "Even our artist with all his skill cannot transmit to paper the flowing metallic lustre visible on these apparently inlaid-with-gold leaves. They are of a rich velvet texture having a metallic lustre and being curiously inlaid with streaks of golden and frosted net work."

Among the creepers which twined their gracious way up the ornate iron pillars of many a stove-conservatory was the cissus, and in *C. discolor* introduced in 1851 by Rollison from Java there was elegant decoration in plenty. The leaves, six inches long and some two and a half broad, are coloured in the "richest manner possible," said the Victorian nurseryman. "It is a rich green clouded with white, peach and dark purplish crimson, the whole covered with a metallic lustre. The underneath of the leaves are a brownish crimson and the tendrils found at the leaf joint are crimson too."

Lapageria rosea or *alba*, with its red or white pendent bells so attractive and self-sufficient in their waxy dignity, contested the bougainvillaea's gorgeous and prolific beauty.

Another Victorian introduction was the dichorisandra from Brazil, *D. mosaica* being a fine specimen with leaves of a lustrous dark green profusely pencilled and veined with zig-zag lines of pure white, the underside being a deep rich purple. The handsome flowers, in a bright azure-blue, are produced in terminal spikes.

A native of Borneo and one of the most beautiful and distinct of plants for the decoration of the stove was *Alocasia metallica* with fleshy shining leaves nearly 2 feet long and some 12 to 16 inches wide, "with an upper surface of a metallic coppery-red lustre so glossy and variable in tint that like the chameleon it changes colour as the spectator looks, according to the light and angle of view—red, blue or purple tints following each other on the bronze surface as the sunshine falls upon the plant, producing an effect which is perfectly gorgeous." To show how they did appreciate a plant they added: "This plant is perfectly unique in its beauty for there is nothing in the whole vegetable world which can rival it."

What joy for the proud owner of the Angel's Trumpet (*Datura suaveolens*) to see in bloom for the first time its almost incredibly large flowers of pure white, near a foot long, sweetly fragrant and so ethereal as to suggest the heavenly instrumentalist with pursed lips ready to blow.

The strelitzia was another ostentatious Victorian with its curiously flamboyant bird-of-paradise flowers, accompanied by that king of stove-conservatories *S. nicolai*, with sword-like leaves 4 feet long and a stem reaching 30 feet to the roof before throwing off its strange inflorescence. A direct link with the great Australian outback were the unusual callistemons, the bottle-brush trees, and for striking novelty in colour and form there were the lobster-like flowers of *Erythrina crista galli* (the coral-tree) or the *Pseudopanax ferox* with hand-shaped leaves; a Dead-Sea-sort of plant if ever there was one!

The artificial curiosities of the grevilleas were other glass-house companions of the Victorians, and to grace innumerable conservatories from India, Borneo, Java, China and the Indian Archipelago came the clerodendrons, the glory-trees, showy scarlet plants, albeit with a graceful air of refinement imparted by their delicate panicles of bloom.

Representing as they did the other side of the world, Trinidad, Brazil, South America, Ecuador and the Philippines, marantas were always a popular Victorian stove-plant, rich, elegant and aristocratic with vivid eccentric markings. Nicholson listed 22 and although it is recorded as being first introduced in the eighteenth century it was not until the nineteenth that it achieved its full popularity.

From 1823 until the end of the century new varieties were constantly being discovered and imported. They might be striped, blotched, dotted or irregularly marked in all manner of tasteful colourings, many of them having the appearance of an artist's palette but lately laid by.

By mid-century it only needed a short tour of the nearest glasshouse establishments to see that floriculture and the growing of foliage plants almost monopolised the glasshouse world. Horticultural authors had caught up and were by now extolling the virtues of this extremely fashionable and exciting cult of hot-house exotics. Fruit was still forced for the table but not quite so much attention was given to it and not quite so much money spent on its cultivation. The money was needed to finance indirectly those hundreds of intrepid plant-hunters who roamed the world to satisfy the insatiable demand for floral and foliage novelty, beauty and rarity.

M'Intosh prefaces his thoroughgoing work on *The Greenhouse, Hot House and Stove* of 1838 with these words: "The cult of exotic plants, whether pursued with the view of producing fruit or flowers, is admitted to hold the highest rank in horticultural science." In 1845—a fair sample year—it was calculated that of the 238 new plant species introduced into this country only 35 were hardy and for cultivation in the outside garden and that all the rest, including 55 new orchids, were greenhouse, conservatory or stove-plants.

James Kain just before the turn of the century could say with truth that floriculture had become the study and amusement of all ranks, and Shirley Hibberd, about the same time, spoke of the increasing attention being paid to stove and conservatory foliage plants, which he claimed was "a distinct phase in the history of horticulture." There was no doubt that at that time a passion had arisen for the collection and cultivation of fine foliage plants and many fine books with excellent colour plates came from the presses about this time to cater for the demand.

Let us now take a stroll into one of the ornamental palm stove-conservatories or winter-gardens of the period. There are the howeas, elegant, zebra-striped with long, graceful cords of fruit hanging sedately down from a crown of feathery leaves. The livistona brings a breath of alien lands with fan-like leaves, some with young petioles and leaves a blood-red, others with dark-green leaves all forming a perfect circle of greenery. Up the

P

STRELITZIA NICOLAI.

PANDANUS ELEGANTISSIMUS.

Two inmates of a Victorian conservatory representing an exotic flora little known at the time.

pillars runs *Monstera deliciosa* with all the tropical attributes of a jungle climber—root twining, large deep-green, largely perforated leaves as if torn by the last hurricane which had passed, as well as large pine-cone-like yellow fruit, pineapple flavoured, some six to eight inches high rising from a nest of leaves.

The cycas over there give a decided oriental flavour to the scene with *revoluta* and *circinalis*, the former massive on its stout seven-foot stem from which branches a crown of feather-shaped leaves anything from two to six feet long, a rich dark green, while *circinalis* is even more opulent with its great feathers in a pendulent crest up to twelve feet long, dark, shining green on the upper side and paler green below.

The chamaerops, too, adds an Eastern flavour with, probably, *humilis*, a handsome columnar stem clothed in rough fibre and the bases of the old leaf stalks giving way to glaucous fan-shaped crowns of foliage.

With *Phoenicophorium sechellarum* (*S. stevensonia grandifolia*) the Victorian grower pointed to what he called the noblest of palms introduced to cultivation, leaves up to eight feet long, four feet wide, fringed at the edges with red, the leaf stems densely armed with black spines nearly four inches long. The blade was deep green spotted with dull orange.

The frangipani (*Plumeria acutifolia*) with curious thick shoots and most sweetly scented yellow and white flowers is redolent of Hindu temples and floral offerings to the gods. From a quite different locale—Peru—comes *Clavija fulgens*, whose tall straight stem bears a cluster of large handsome leaves and deep-orange-red flowers waxy and starry in their bizarre beauty. The carludovicas throw up their fourteen-foot-high fans ready to temper the tropic air to any Eastern princeling. Huge heads of flaming orange-scarlet, near three feet across, rivet attention, flowering as they do from the branch ends of the tree we have come to now; it is *Brownea macrophylla*, from Central America. The young leaves of this tree, it is pointed out to us, hang wiltingly down, a brownish grey colour, yet shortly, as they age, they will turn green, stiffen and assume their normal pattern.

These tropical flowering trees, ever since Chatsworth had flowered *Amherstia nobilis* under glass earlier in the century, always attracted the Victorians and in this collection is *Saraca indica*, appealing flamboyantly with large dense clusters of bright orange flowers later turning to red. This too has a novel leaf habit, the young growth being red and hanging vertically down. Adding to the graceful weeping effect in the house is *Grias cauliflora*, a tall slender unbranched tree terminating in a fine crown of drooping glossy green leaves as much as three feet long.

Ferns of all kinds fill in the interstices of this home-made jungle, osmunda, aspleniums, dicksonias, cibotiums and angiopteris among others, carpeting the forest floor with a bewitching greenery exotic in its lushness and its extreme grace of form.

A different sort of foliage occurs in chrysalidocarpus, the feather palm from Madagascar, a graceful slender stem with ringed bark running up to ten feet or so before throwing off its crown of delicate fronds; and, decorative indeed, nearby is *Sabal blackburniana* from Bermuda with a magnificent sheaf of foliage of immense size with the divided leaf margins a blue-green and at the point of attachment to the leaf stem a great triangular blotch of yellowish white coloration.

The cocos family brings a breath of South Sea Island magic into the already foreign atmosphere. Those tallest trees with the great crests are the cocos, probably *plumosa*, always a favourite in the larger houses, highly ornamental with long arching leaves and drooping bunches of waxy flowers followed by the orange-coloured edible nut. Useful to ring the colour changes is thrinax or the silver thatch palm, like *argentea* with leaves of silvery grey and a growth up to twelve feet.

A tree brought in to add variety in leaf form, yet still a palm, is the graceful *Caryota cumingii* from the Philippines with slender trunk bearing a spreading crown of foliage up to six feet in length and three feet wide, but with cuneiform leaves.

The aborescent ferns so characteristic of tropical flora are represented by the cyatheas and a quite remarkable one here is *medularis* with fronds resembling iron bars, as regularly arranged as the arms of a giant candelabra, sixteen feet long.

Calathea veitchiana nearby introduces a new note altogether, with leaves representing, so the owner says, peacocks feathers in green, yellow, white and olive on the upper surfaces and rosy pink beneath.

Pandanus or the screw-pines are very much a feature of these palm-conservatories and tropical plant-houses with their three feet long ascending leaves, green, white margined, red spiked and strangely angular.

The corypha family produces the largest leaf fans of any of the palms and from stout tall trunks, those we are under, throw gigantic crowns of immense fan shaped leaves so long and strong that natives of India and Ceylon use them for thatching, while over there *umbraculifera's* massive six-foot leaf-stalks support large fan-like leaves which are plaited and form a complete circle some twelve feet or more in diameter, a ready-made umbrella.

Ornamental and decidedly different are the xanthorrhoea, the blackboys or grass-trees from Australia with stout stems which from a height of six to eighteen feet throw down a great mass of grass-like leaves, like a great untidy antediluvian bird's-nest.

The tropical bamboos and calamus, the climbing jungle palm canes, clothe the rafters and pillars to add verisimilitude to the jungle picture with which we are presented, and of course the musas, the bananas, with their huge dark-olive pendent leaves never fail to give the finishing touch to this tropical scene.

17

A JUNGLE POOL

In a nearby house connected by a glass corridor ranged with the vase-like bromelia in all their striking forms and exciting colorations, is the Aquatic House—but we might well spare a moment for these epiphytes of the forests of tropical America around us before we pass on.

These so solid, dark-green and blood-red water chalices, formed of tightly wrapped rosulate strap-like leaves, are alien and unreal, with such forms as *Aechmea fulgens* which produces brilliant scarlet bracts and blue flowers, or *distichantha* which from a matted crown of long glaucous leaves tapering to a point and armed with reddish-brown spines, throws up a flower spike densely clothed with bright-red bracts and blooms of rose and bright purple. All the varieties astonish by the contrast of brilliant bracts and flowers of stabbing deep blue, purple, pink or bright red, rising from such stiff, tightly-wound bases of leathery foliage.

Caraguata ʒahnii, another South American, is showy with yellow flowers above intense scarlet bracts, and yellow leaves with scarlet stripes, the whole of a semi-transparent nature.

The tillandsias are a noble plant family, and grown here both for foliage and floral beauty are *splendens*, and *lindeniania*, the latter having leaves arranged in a rose-like manner, tapering up to end in a fine point, light green on the upper side, suffused rose under, all marked with parallel lines of reddish brown; the scape being a rosy carmine and the flowers—probably the most elegant of the collection—azure with the purest of white eyes.

Splendens near by boasts fiery purple bracts, nosegaying yellow flowers set amidst a most graceful leaf formation, and, as if rounding off the show, there are the ananas, the pineapples so well known and so well grown in British gardens for many years previously. Here the variegated-leaf forms are grown for show.

Giant Water lily.

In the aquatic house near by water and heat form a pleasant tropical atmosphere in the English mid-winter and we can imagine its Victorian owner ambling along the winding paths walled with jungle verdure, wearing topee and immaculate white ducks with no sense of incongruity at all, as he enjoyed a cigar on a post-prandial stroll!

In such an atmosphere vegetation is lush, prolific and Brobdingnagian; nature has seemingly run riot, rainbow hues are matched and surpassed, greenery thrusts against greenery, yet the gentle splash of water over a fern-decked fall gives a peace most satisfying to the harassed Victorian paterfamilias.

Temperatures run anywhere from 68° F. to 95° F. and the central feature of the house is, naturally, the great water-tank disguised with rockwork, skilfully planted borders and artificially placed climbers, so that it bears every appearance of having been discovered entire in some jungle clearing and transported in one piece to this English home. Winding paths, in and out of the greenery, skirt little hills, minute lawns of *Lycopodium denti-*

culatum open out the delightful scene and a rivulet runs between a valley of selaginellas. Along the paths, lichen-covered tree-trunks, as natural as artifice can make them, bear their epiphytal fernery ornamentations.

The owner is fortunate in having as a central attraction of his house the vegetable wonder of the Victorian world, *Victoria amazonica*, the giant Amazonian water lily, whose large circular tray-like leaf, five to six feet across, will hold a child upon its surface as upon a raft, and when it flowers a foot-wide satiny white mass of petals floats its magnificence around a great button-centre of rosy purple. All around are the floating elegances of the nymphaea, to many of their growers the ultimate example of tropic flora. Their range of colour, their sweetly smelling chalice blooms add a rare richness to the water scene and the surrounding verdant foliage. Under artificial light, and many were the conservatories and plant-houses lit with gaslight in artistically ornamented lanterns, the colours glowed with a richness breathtaking in their fierceness and translucent glitter.

Nymphaea lotus dentata, for instance, has a pure white chalice of a flower with massive leaves, or *gigantea* a beautiful blue lily had a dense mass of deep golden stamens quite seven inches across; the great shield-like dark-green leaves were upwards of two feet in diameter.

N. rubra, night-flowering from India, brings a splash of the Orient with its 10-inch bright-red flowers and cinnabar stamens contrasting so strongly with the bronze, crimson and green foliage; *devoniensis*, raised at Chatsworth in 1851, a hybrid of *rubra*, has even larger cups, bright rosy red they are, a foot across, and its foliage is bronze-red colour. *N. amazonum* has graceful oval-shaped leaves, bright green, with large yellowish-white clustering flowers with a sweet odour suggestive of peaches.

N. caerulea holds its sky-blue flowers with black-spotted sepals well out of the water to dispense its sweet fragrance from 7.30 p.m. to midnight.

N. capensis, the Cape lily, also bears a lovely sweet-scented flower of a rich sky-blue with green, blotched with purple, leaves.

It was always difficult to choose a favourite among so many regal blooms but *N. zanzibariensis* was a beauty among beauties with its profuse flowering blooms of rich midnight-blue, fragrant, and each golden stamen tipped with navy-blue.

The shield-like leaves of *Nelumbo nucifero* are borne on tall stems with the tender rosy-pink flowers shading to delicious orange most delicately scented, also held well out of the water and seen to the best advantage at breakfast-time.

Pistia stratiotes, the tropical duck weed, quilts the pool with delicate pea-green verdure and *Ouvirandra fenestralis* from Madagascar, with lace-like leaves, never fails to attract in the cool shallows. A profusion of white flowers, yellow-centred, rising from small heart-shaped leaves betokens the water-snowflake (limnanthemum) and is always a favourite.

Gracing the shores and intriguing miniature bays of the pool are the tropical bamboos, stabbingly linear after the curvilinear grace of the nymphae foliage. *Limnocharis humboldtii* is a handsome ornament, its smooth oval leaves, bright green in colour float near the edge by bright yellow cups. The great circular leaves and the flowers of *Nelumbo speciosum,* the sacred bean of Egypt, rise on slender four-foot stems from the water's edge, the flowers being a delicate rose and white and heavily fragrant.

Apanogeton in its varieties brings diversity to the satisfying foliar geometric patterns laid out on the waters with *dinteri* and *leptostachyus abyssinicus,* the former with oblong floating leaves curved at the base and two-spiked inflorescences of yellow flowers, while the latter has twin spikes of lilac flowers and small strap-like floating foliage.

Less striking in their forms and colours, but all adding charm and contrast are the water hyacinths—*Eichhornia azurea* from whose shining thick leathery leaves a stout stalk rises to hold funnel-shaped bright lavender-blue flowers; *martiana* which with heart-shaped leaves below has flowers of rich purple, lemon-yellow blotched; *Limnobium bogotensis* with small round deep-green shining leaves; and the elegant fern-like *Salvinia natans.*

Around the edges of this highly decorative water, adding

exotically to the tropic make-believe, are the colocasias whose elephant-ear leaves are a dull green purpling metallically up for three feet and the paper-plant *Cyperus papyrus* and *diffusus* whose mop-heads on tall slender stems add a willow-like charm to the scene.

The cannas add fire and sword-like leaves to the picture while the large orb-like foliage and massive yellow-white and various shades of rose of the nelumbiums add weight and dignity. *Xanthosoma lindenii's* large arrow-shaped leaves a foot long are distinct with their dark-green background picked out by the white of veins and midriff.

Anthuriums put out their fiery tongues and *Hedychium coronarium* gives incense from strongly scented yellow flowers streaked with red. *Acalypha hispida* casts down its long blood-red cat's-tails to reflect in the water from which, struggling up from the warm mud, are the lagenarias, the gourds, striking objects with brightly coloured and weirdly shaped fruits; one there is the loofah plant. In all their flaming beauty the hibiscus and the bougainvillaea struggle with the less fiery character of *Ipomoea digitala*, that profuse tropical creeper with the pinkish-mauve flowers, and *Clerodendrum thomsanae*, all weaving a fantastic curtain of rainbow patchwork above the pool and clothing the moss and fern-covered rockwork walls with a torrent of hues no canvas can capture.

Adding to the tropical atmosphere are the arums, the alocasias with bizarre colouring on great boldly presented leaves, and some of the more colourful of the nepenthes.

18

ORCHIDACEOUS BEAUTY

BEAUTIFUL, rich and varied as the collections were, none of the plants we have just described could compare in wondrous enchantment and aesthetic appeal with the orchid.

Not grown at all in this country, except for an odd plant at Kew, at the end of the eighteenth century, within forty years thousands upon thousands of plants were to be found in the stoves and greenhouses of Britain. So soon as the forests and hills had revealed their exotic treasures to the horticultural world an orchidomania arose to outvie the tulipomania of a previous century. Single plants changed hands for as much as £1,000 and quite regularly in the early days anything from £100 to £500 was paid for a piece of a rare specimen in the famous London auction rooms. Collectors were sent scurrying to all known haunts of this bizarre jewel of nature for, to the wealthy gardening Victorian, this plant had everything—strange unnatural forms, heavy, quite often overpowering fragrance, previously unheard-of colourings and markings, curious habits of growth and foliage and, singularly romantic, nay, legendary stories, behind its collection and discovery.

Like so many other exotic plant introductions to this country the orchid came at the right time for its proper keeping and cultivation. The few in the country at Kew from 1800 up to 1820 were never in the best of health, and it was not until Mr. William Cattley of Herefordshire about that time, by hit-and-miss experiments, finally discovered the best methods of orchid-cultivation that an orchid collection became a practical proposition. His methods were copied and once more the stage was set for the curtain to rise on another great English glasshouse tradition.

Just what Mr. Cattley had started is indicated by a statement of

Lady Amherst, who said in 1896 that nearly every single orchid in the country had been imported during the last fifty years and that most of their known haunts had been ransacked. During one particular search for *Odontoglossum crispum*, 10,000 plants were found and collected and in the process 4,000 forest trees were hurled to the ground. The natives would not climb so the trees had to come down!

This was a far cry indeed from the early orchids at Kew, and a floral revolution compared with 1763 when Linnaeus had first told incredulous horticulturists about the existence of the orchid and had sought to prove his point by showing them herbarium specimens and some drawings.

As had been noted previously Peter Collinson in 1731 is recorded as having received the first exotic orchid and this was followed in orchid history by the flowering of *Phaius grandifolius* in the stove of Mrs. Hird at Apperley Bridge, Yorkshire. She was a niece of Dr. John Fothergill who had the plant brought from China. In 1787 *Epidendrum cochleatum* flowered at Kew, but importation was in ones and twos and experience was inadequate to handle them. Probably one of the earliest growers to have any success, no doubt contemporaneous with Mr. Cattley or even imitative of him, was the Earl Fitzwilliam at Wentworth Woodhouse in Yorkshire, and the Chatsworth collection began soon after in 1833.

Once those early fumbling attempts at cultivation had been turned into supremely successful ones there was every reason why the orchid should become the florist fancy of the age, the hallmark and symbol of Victorian glasshouse supremacy. As Frederick Boyle, a writer on orchids, said of the plant at the beginning of the nineteenth century, "it was expressly designed to comfort the elect of human beings in this age." He claimed that their growth "must needs have been the latest act of creation in the realm of plants and flowers" and that until that time—mid-Victorian—civilised man could not have in any way enjoyed the flower. Until the end of the eighteenth century ships were not fast enough to convey them alive so that "the blessing was withheld from civilised man until, step by step, he gained the conditions necessary to receive it. Order and commerce in the first place; mechanical invention next, such as swift ships and easy communications; glasshouses and a means of heating them which could be regulated with precision and maintained with no excessive care; knowledge both scientific and practical; the enthusiasm of wealthy men, the thoughtful and patient labour of skilled servants—all these were needed to secure for us the delights of orchid culture."

Mr. Boyle certainly made out a most accurate case for his enthusiasm—orchids—but it was the same picture, the same conditions which were necessary, as we have sought to show, before the glasshouses of this country generally could become the glorious show-houses they were in the Victorian and Edwardian eras.

Probably some of the legends of orchid-hunting, of orchid-growing, some of the *obiter dicta* which so quickly grew up around this glorious plant will give a glimpse into a horticultural past we are not likely to experience ever again.

There is the amazing story of *Cattleya skinnerii* from Costa Rica, the largest single specimen ever discovered—7 feet in diameter it was and 6 feet high. In full bloom it held fifteen hundred orchid blossoms the scent of which could be enjoyed over a hundred yards away. This gargantuan piece of natural beauty was found growing in the fork of an immense forest tree. The native tribe in whose territory it was, almost worshipped this floral giant and refused all offers for its removal. It was not until the famous St. Albans firm of orchid-growers made the bait overwhelmingly tempting that they finally gave in. Then the whole tremendous plant with the tree-fork it grew in, weighing altogether 1,200 pounds, was cut down and man-handled hundreds of miles through rough jungle country to the port and thence to Southampton and St. Albans where a new house was built to receive the treasure. Here the plant and tree-fork were suspended from the roof by chains and thousands flocked to wonder and gaze.

In the spring of 1894 one firm alone, Messrs. Sanders, had

twenty collectors searching the jungles, with two in Brazil, two in Colombia, two in Peru and Ecuador, one in Mexico, one in Madagascar, one in New Guinea and three in India, Burma and Straits Settlement. So tremendous was the demand, so keen the collectors and their firms to meet it that some orchid-growing areas of the world were devastated. Indeed, had the demand continued at the peak rate of the last half of the nineteenth century then an orchid close season would have had to be declared to give new stock an opportunity to rise from the sites of jungle carnage. As it was, most successful hybridising made up for the new discoveries which were not forthcoming.

Of *Laelia elegans* when it was discovered in 1847 it was said that the small island off Brazil where it grew was so full of the plant that no such orchid horde had ever before been found, yet less than fifty years later not an orchid remained on the island.

Millican in 1891 complained that "most of the orchid-growing portions of the world have been ransacked and their glorious plants packed over by thousands to this country and in some cases their natural habitat left bare."

Whole districts were denuded of their forest trees to get at the epiphytal orchids twining their graceful flower-spurs in the upper branches. The woods which formed the home of *Miltonia vexillaria* were said to have been cleared as if by forest fire, and during one particular search for *Odontoglossum crispum* in Colombia, when 10,000 plants were collected, as has already been mentioned, 4,000 trees were cut down; the camp of the collector being moved forward week by week as the trees were felled and their plant wealth exhausted.

Roezl, one of the great collectors of the century and a name to be conjured with in the annals of the plant, on one occasion forwarded eight tons—yes, eight tons—of orchids to Europe as the result of one of his very many and very successful plant-hunting expeditions.

Something of the scale of these orchid-butchering jobs can be gauged from a letter to England in 1895 from Carl Johannsen writing from Medellin.

"I shall despatch tomorrow," he wrote, "30 boxes, 12 of which contain the finest of all the Aureas, the Monte coromes form, and 18 cases containing the great Sanderiana type all collected from the spot where they grow mixed, and I shall clear them out. They are now extinguished in this spot and this will surely be the last season. I have finished all along the Rio Dagua where there are no plants left; the last days I remained in that spot the people brought in two or three plants at a time and some came back without a single plant."

In the auction rooms stakes were high and spectacular bidding took place for the newly discovered floral gems. Quite often the auctioneer was able to add to his professional patter romantic tales of daring as new, rare plants came up to the rostrum. For instance, no orchid enthusiast ever forgot the day they sold *Dendrobium schroederi* at Protheroe's London rooms. A unique occasion indeed, for the plant had been much sought after and there it was at last; but if you bought the plant you also bought the human skull it was attached to. The natives of the island of "the Australasian seas" where the orchid was discovered would not allow their beloved icon, ethereal companion to their tribal idols, to leave their keeping except on two conditions, one of their idols must accompany the flower, and secondly the flower must not be detached from the human skull on which it had entwined its roots so lovingly. To do otherwise, they said, would bring disaster upon the tribe, and thus it was that in a staid Victorian auction room a human skull and a native graveyard idol accompanied a beautiful flower put up for public sale.

Not only were the collectors avid for all they could find, but such was the insatiable demand of the market that hybridising and propagation also went along at a terrific rate to match the collector's pace. Of odontoglossums, for instance, it is recorded that in 1883 there were but five known species, yet so fast was the plant discovered, hybridised and propagated that at the turn of the century one British nurseryman, Messrs. Shuttleworths of Clapham, carried a stock of 10,000 in his greenhouses.

In terms of hard cash the trade and the amateur grower

represented thousands upon thousands of pounds in plant value as is well shown by the sale about the end of the nineteenth century of Mr. Day's stock of orchids from his Tottenham nurseries which brought in £24,000 from an American syndicate.

No, there was no question, no question at all that the cult of the orchid went hand in hand with the possession of wealth; indeed the flower became and still remains a Croesus symbol. However it was not solely a question of a man's ability to pay for it, for there was a vital and healthy spirit of competition between the extraordinarily enthusiastic amateur growers, so magnificently backed by their professional gardeners, and they vied with each other to have the latest and the best. They really loved their flowers, too, and waxed lyrical about them: *Cattleya aurea imschootiana*, they wrote, has a splendour beyond conception, the flowers 9 inches across and the colour of the sepals and petals mauve with a crimson-purple lip, and a golden throat lined with bright crimson.

Of *Cattleya sanderiana* they enthused: Here is a flower 11 inches across of a tender rosy-mauve. The vast lip is almost square with a throat of gold lined and netted over with bright crimson. It has the charming eye of *gigas* in perfection and the enormous disc is purple frilled with the liveliest magenta crimson.

A whole literature grew up around the orchid family and some of the finest colour-illustrated books in horticulture were published to satisfy the love for the plant and the hunger for its life history in detail. It may be remembered that Darwin played his part in the orchid story. When he saw *Aeranthus sesquipedale* with its narrow throat 14 inches and more in length he postulated that somewhere in the jungles of Madagascar there must of necessity be an insect, probably a moth, with a proboscis 14 inches long, which could reach to the nectar held by the long, narrow storehouse and in doing so trigger off the fertilising machinery. Although no one had ever heard of such an insect, Darwin was later proved to be right.

Is there little wonder that the orchid was considered to be one of the marvels of the nineteenth century?

19

DECLINE AND FALL

THE twentieth century brought with it a revulsion of taste from the fussily ornate and the brashly ostentatious towards the more natural and the simple, in all aspects of life, including the garden.

The levelling process that had begun in the nineteenth century soon got into full swing, narrowing the gap between rich and poor and, whilst it involved the decline of the great estates and country houses, effected a great improvement in the living conditions of the masses. In many ways the indulging of individual tastes, the enjoying of a hobby, which for hundreds of years had been the pleasure of the rich alone, the poor having neither the time, the inclination nor the money, was now the pleasure of millions. The cheap newspapers, the flood of educative, reasonably priced books on every subject under the sun for a populace which not only could read but now had leisure, had brought that about. And gardening in the glasshouse, once the prerogative of the rich, now found a broader base on which to operate among many of the less wealthy.

The loss of much of their agricultural wealth resulted in a reaching out for investments by the upper classes in the up-and-coming industries of the town, so that estates were now being run on the backs of factory-hands not on those of hinds, gardeners and agricultural labourers. This inevitably broke the close association between the aristocrat and the land, including the huge estate gardens and glasshouse establishments.

High labour and material costs made maintenance a drain generally, and when maintenance was required for ironwork and glass conservatories or the huge wood and glass ranges which the Victorians had built without thought of expense, then the costs were high indeed. Many an owner found it cost more to cut new intricately shaped panes for the hundreds of feet of ornamental

iron-scrolled trellised work than it did to have the whole thing
pulled down or altered to a less ornate style which made straightforward the job of glazing and maintenance.

By the end of the nineteenth century it was already quite
apparent that the gardening scene had changed, and that the
pendulum had swung again. There was a distinct breakaway from
the floridness of the conservatory to a simple hardy herbaceous
plant, shrub, tree and alpine garden scene.

Collectors no longer dragged their expeditions through steaming jungles and fetid tropic swamp, they climbed higher to rarer
airs for hardier plants, and, with Forrest, Wilson, Purdom, Farrer
and Kingdom Ward, a new race of plant-collectors was born to
match the patterns of a new century and a new century's gardens.
Now it was the Tibetan heights, the Chinese mountain regions
and the high barriers on the roof of the Asiatic world which were
the hunting-grounds for new plants for British gardens.

William Robinson, in his *The English Flower Garden* which,
first published in November, 1883, ran through eight editions and
six reprints up to June, 1903, was the first to thunder in trenchant
and indeed bombastic argument for a return to Nature, for a
complete switch from the costly artificialities of the hot-house
and conservatory to the simple, the hardy and the comparatively
cheap. He appealed to good taste and to tradition for a return to
the old-fashioned cottage garden, and with great effect. Many
authorities date the start of the glasshouse decline from Robinson's theories and practice. It should also be considered, however,
that his writings were well-timed, either by accident or design, to
appeal to a changing age and fashion.

Robinson, an Irishman, started his gardening life as he meant
to go on, with an act of sabotage which qualified him for the role
of the glasshouse iconoclast of the century. As a clever young
man of twenty-one he was put in charge of a range of greenhouses
on the big estate of an Irish clergyman-baronet, where he successfully tended and cared for a fine array of tender and rare plants,
until one raw winter's night in 1861 when, after some squabble
with his employer and with the thermometer below freezing, he

drew all the fires, opened all the windows and left his precious charges to freeze slowly to death.

Naturally Robinson did not stay to see what effect this might have on his employer. He left immediately and soon after had got himself employment as a foreman at the Royal Botanic Society's gardens in London. He prospered exceedingly, but never failed to rail in letter, in books and journalism at bedding-out and greenhouses. In his great work, *The English Flower Garden*, he wrote:

"Among the evils of the 'bedding' and 'carpet system' is the need of costly glass houses in which to keep the plants all the winter, not one in ten of these plants being as pretty as flowers that are as hardy as the grass in the field—like roses, carnations and delphiniums. It is absurd to grow alternantheras in costly hot houses and not to give a place to flowers that endure cold as well as lilies of the valley.

"Glass houses are useful helps for many purposes, but we may have noble flower gardens without them. To bloom the rose and carnation in midwinter, to ripen fruits that will not mature in our climate, to enable us to see many fair flowers of the tropics —for these purposes glass houses are a precious gain; but for a beautiful flower garden they are almost needless and numerous glass houses in our gardens may be turned to better use. It would not be true to say that good hardy-flower-gardening is cheaper than growing the half-hardy plants that often disgrace our gardens, as the splendid variety of beautiful hardy plants tempts one to buy; and it is therefore all the more necessary not to waste money in stupid ways, apart from the heavy initial cost and ceaseless costly labour of the glass house system of flower decoration.

"For those who think of the beauty in our gardens and home landscapes the placing of a glass house in the flower garden or pleasure ground is a serious matter, and some of the most interesting places in the country are defaced in this way. In the various dividing lines about a country house there can be no difficulty in finding a site for glass houses where they cannot injure the view. There is no reason for placing the glass house in front of a beautiful old house, where its colour mars the prospect though often, in looking across the land towards an old house, we see first the glare of an ugly glass shed. If this were the case only in the gardens of people lately emerged from the town to the suburbs of our great cities, it would not be so notable; but many large country places are disfigured in this way. And apart from fine old houses and the landscape being defaced by the hard lines and colour of the glass house, there is the result on the flower garden itself; efforts to get plants into harmonious and beautiful relations are much increased if we have a horror in the way of glass sheds staring at us. Apart from the heavy cost of coal or coke, the smoke-defilement of many a pretty garden by the ugly vomit of these needless chimneys; the effect on young gardeners in leading them to despise the far more healthy and profitable labours of the open garden; all these have to be considered in relation to the cost, care and ugliness of the glass nursery as an annual preparation for plants for the flower garden, these plants being with few exceptions far less precious in every way for flower garden, or for rooms than those that are quite hardy.

"A few years ago, before the true flower garden began to get a place in men's minds, many of the young gardeners refused to work in places where there was no glass. A horrid race this pot and kettle idea of a garden would have led to; men to get chills if their gloves were not aired. I met the difficulty myself by abolishing glass altogether. Only, where we do this we must show better things in the open-air garden than ever flourished in a glass house."

That was a tirade if you like. Robinson was quite adamant about glass and was as good as his word at his beautiful home and the garden he landscaped at Gravetye Manor in Sussex where he tore out every pane of glass and rased the houses to the ground.

This was plain speaking backed by action; indeed the whole book in its extended argument for wild gardens, colourful hardy herbaceous borders, rockeries and the softening all round of the rather harsh lines of garden formality as practised at the time as a clarion call and a rallying point for the many forces tending

to weaken the stranglehold of bedding-out greenhouses and conservatory on country-house gardening.

The clever plantsmen in this country also played a part, and
by no means a small one, in the revolution. Their willingness
to experiment had proved that many plants thought by all when
they were first introduced to be precious hot-house subjects to be
coddled and cosseted, would live with less heat without damage,
so that the cool greenhouse began to be the repository of scores
of plants which had not, until then, been trusted out of their
Turkish bath atmospheres. Some even found their way into the
sheltered borders.

They had, too, worked wonders in hybridisation. Even if more
plants for the few owners who managed to keep going stove and
conservatories had been available in their native haunts (so
thoroughly had much appropriate territory been combed), the
hybridisers had been so successful that it is doubtful indeed if
better indigenous plants would have been found than those bred
in the nursery establishments of both professional and amateur
growers. It is only just over a hundred years ago that the first
hybrid orchid to be produced in Britain bloomed at St. Albans,
being the precursor not only of a great industry but of some
amazingly beautiful and bizarre flowers.

All manner of other factors were also militating against the
upkeep of splendid glass ranges, Tennyson's "Tropic Squares."
Men generally, except the few remaining scions of the noble
houses with some money still to burn, had less leisure, with the
result that in the long week-end in which meals, routs and dances
in the magnificently bedecked conservatories had always been
such an attractive feature went by the board. Men were drawn
more into the actual workings of their offices, business or pro-
fessions; the increasing competition from Germany and the
United States in so many fields made wealth less easy to come by;
it had now to be earned. Women, too, with the expanding world
about them had extensively widened interests and their days were
not now taken up with seeing to the flowers, which had been
almost a full-time job for the mid-Victorian mama and her

daughters; thus the woman of the house's direct association and link with the garden and the decorative potentialities of greenhouse and conservatory was greatly weakened.

Men who could once be got for a mere pittance to tie themselves heart, soul and body to an estate and to its gardens were becoming harder to find; the cities and towns called, where wages were higher and life was infinitely more varied, while immigration took many others, who were willing to try out life in a new world.

There was a "life is real, life is earnest" attitude which had to be taken by most people, and in the changed way of life selfglorification or self-indulgence in the costly hobby of a large glass establishment had no place, nor was there room for it within the tighter household budgets at the turn of the century.

There was more to do for people who loved a garden and a greenhouse too, now that it was possible to travel easily from place to place and now that public entertainment had become a thriving business. All manner of fruit shipped from all parts of the world in quantity was to be had in excellent condition and cheaply at any corner shop, which made a sorry economic mess of a fruit-house range. And flowers and pot-plants, not only from our own commercial nurseries which had sprung up to meet a wider public demand for the more ordinary floral products, but from the Dutch markets, were also easily and economically available.

It is to be regretted, but with the antimacassars, lustres and wax fruit under glass of Victorian England, went the fern cases, the terrace-house conservatories, practically all the highly ornamental glass, the rococo conservatories, the rockwork greenhouses, the palm-decked tropic stoves and the jungle-like aquatic house. It was a natural clearing away of what came to be felt was ostentatious, unnatural knick-knackery, in a utility world with little to spare for fripperies.

Also, apart from the very few show-places of the land where tradition bade the incumbent keep up the showlike covered garden tradition with luxurious floral and plant displays, albeit in a reduced state to which the family had been accustomed, by the beginning of the twentieth century the middle classes had become the influential majority and they were definitely practical no-nonsense types who got on with the work at hand, and lived largely in towns and cities. They had their share of greenery and blossom in the big public parks which sprang up in the vicinity of their long-terraced homes.

So it was that when the first war came with its call on young men, with its limitation on fuel, with the vital necessity for home-grown food, not fruit, but down-to-earth potatoes, cabbages and peas, the one-time premier world position of the British glass-house tradition had all but gone. The war years themselves dealt it a body-blow from which it has never recovered its original magnificence and scope.

Between the wars what recovery took place was in an entirely different direction from the walled gardens of the great houses and the costly glass ranges available for all purposes; it was in the world of the small men, the Mr. Pollys with their allotments and little suburban handkerchief gardens, into which they delighted to squeeze their 6 × 8s for an annual crop, God willing, of chrysanthemums and tomatoes.

The 27 May, 1920, might well have been chosen as the symbolic day when the Victorian Tradition died for ever from this land, for on that day the fall of massive wooden beams and masonry and the sharp crack and reverberations of explosive charges surely dealt the glasshouse cult a death-blow as the Great Conservatory at Chatsworth was rased to the ground.

A contemporary writer thus describes the scene and sets the seal on an era:

"The far-famed beauty of the grounds of Chatsworth House, the noble residence of the Duke of Devonshire, contained no more striking adornment than the great conservatory. This im-mense glass-covered structure has inspired wonder and admiration in the hearts of countless thousands of visitors to whom the opportunity of inspecting the Chatsworth demesne has been generously offered by successive dukes. It was notable, not merely because of its great proportions, but also since within its light-giving exterior was probably the finest and most representative collection of tropical plants that this country knew. Here one could see the palm and the banana trees growing as if indigenous and not exiles. Delicate ferns and minute aquatic plants were preserved in their full richness. Here nature-lovers came in a mood of sedate delight, and even the casual visitor, one of a jolly party maybe, was struck with awe at the quiet majesty of the place and its rare tenants.

"I stood in the rain today, in a tree-girt enclosure, looking long and sadly at a dismal expanse of debris stretching away from my feet to where tall trees swayed down as if to hide the spectacle. Great iron pillars snapped in several pieces littered the ground. Thick baulks of timber, split and shivered, sprawled about. Over the turf was spread a glittering carpet of broken glass, some of which has frosted the long strips of iron lattice work included in this motley company. A stone ornamented doorway stood isolated at one end of the mass. At intervals were gaunt iron columns, apparently firmly erect, and along one side of the enclosure was a curving glass screen, extending the entire length of the open space, and its top was jagged. Out of the litter two withered palm trees raised their forlorn heads; a cluster of ferns sprang freshly-green from a giant stalk. But elsewhere the falling beams and columns had borne down into the earth the numerous trees and plants. My companion, an old retainer, exclaimed, 'A national calamity!' I agreed, for this vast heap of broken glass and metal was all that remained of the proud conservatory of Chatsworth.

"The conservatory may be described as a victim of the war. When the days of stringency arrived, it was found that the preservation of tropical flora was not a work of 'national importance.' Coal was not allowed to be used in keeping up the necessary temperature. There were twelve miles of six-inch pipe coiled about the interior, and the fires beneath the building consumed about 350 tons of fuel a year. The conservatory grew cold as the winter of 1916 was entered upon, and the green things began to fade. Probably the following summer would have restored them, had there not been the sting of a severe frost in the tail of the departing season. The conservatory became a house of death. In the conditions which followed the close of the war, the task of collecting new specimens became undesirable. The cost of up-keep was heavy, and taxation grew no lighter. With the reluctant approval of the Duke, who was in Canada, it was determined that the conservatory should be removed and sold.

"The building upon which sentence was passed was built by Sir Joseph Paxton (then Mr.) about eighty years ago. It was designed in the form of a trefoil, and erected on what is known as the 'ridge and furrow' system, with iron ribs. Its bulk was supported by 48 iron pillars, each of which weighed three tons. The interior was 277 feet long and 123 feet wide, and the centre dome soared 67 feet above the ground. Seventy thousand square feet of glass went to make the sides and roof. The middle aisle was wide enough for the passage of a carriage, as was shown on the occasion of a visit by Queen Victoria in 1843. A staircase hidden among the rockery gave access to a gallery which ran around the place. The Crystal Palace, it is stated, was built after the same principle.

"Preparations for the work of demolition have been proceeding for some time. A quantity of the glass in the more accessible parts of the conservatory was removed. Afterwards charges of gelignite were placed under several of the pillars. Early on Monday morning they were detonated. There was inevitably a noisy explosion, and a few of the panes were blown out. But the towering structure appeared little distressed. Further charges were laid, and yesterday morning, as the newly risen sun was glinting on the expanse of crystal, a series of explosions occurred. The whole conservatory seemed to sway, then steadied itself as if in dignified protest against modern methods and exigencies. A few minutes later the glass in the domes began to screech and crackle, the rending of wood and iron was heard, and down toppled the massive beams and pieces of metal. The shattered glass fell more slowly, the minute fragments descending on the debris like tears. More explosions will be needed before the whole building is down. Even yet, 22 pillars refuse to fall. Assuredly the conservatory was built to endure the stress of time if not 'the slings and arrows of outrageous fortune.' Two months more, and there will be nothing to show that in this part of the grounds there stood what was aptly described as 'the wonder of Chatsworth.' "

What a come-down was the new era, what a poverty-stricken reminder of the glories that once there were! Yet what this decade of gardeners lacked in colour, diversity of form and richness of collections they made up for by an almost pitiable keenness, pitiable in that it was geared to such a miserable aim.

It was the little men in their millions who kept the tradition ticking over until that other great tragedy, World War II, which disposed not only of the little man's efforts but of the last vestiges of any long-lingering Victorian glasshouse splendour.

In both the wars vast horticultural riches were lost, many never to return. During the war years there was no heat to spare for the resplendent or the horticulturally curious, and on to the compost heap or on to end-of-the-garden fires went the spoils of three hundred years, of a thousand intrepid hunters, and the hardly-won floral magnificence of a tropical world.

When peace returned after the last conflict, however, all was not lost. It was still a world of little men, but more and more of them became the proud owners of a little greenhouse—the score is estimated at between one and a half and two million small greenhouse-owners excluding commercial glass now standing at some 436,000 acres for all purposes!

20

THE GLASSHOUSE AND
THE FUTURE

WHILE much was lost in plant wealth much was gained in wisdom and knowledge, for although it would be exceedingly difficult if not impossible to fill a stove or conservatory today to match in any way a Victorian establishment in its heyday, knowledge as to growing, handling, propagation, potting, and everyday maintenance techniques for plant life is better known and proof-tested than ever before in history. The scene has switched completely now from the splendidly showy to the prosaically precise.

It is probably true that never have so many gardeners been greenhouse-owners as at the present time, but it is also true to say that the greater majority of these are small-house amateurs and amateurs who have little enough experience of even the commonest plants, although their technical skills and techniques in running those houses have never been more easy to acquire or to carry out.

It has always been accepted that the growing of greenhouse plants from widely separated parts of the world, with widely differing temperatures, of both soil and air, atmospheric pressures, light intensities, and daylight hours, could be but a compromise at best.

The old gardeners knew of these difficulties and often wrote of them—the wonder was they overcame them at all. One of them spoke of simulating natural conditions for the new and foreign plants, but so that the plants would not suspect the trick being played on them. He said it was necessary to introduce the imposed climate by stealth, bringing on in turn first spring, then summer, making the while artificial dews and showers. They knew that the culture of plants in an artificial climate was a very chancy affair and that only by strategy could they simulate the native natural conditions of any newcomer. They and their masters realised that the greenhouse was a work of art and that the plants in them required constant and unremitting care and attention to "counteract the tendency of that artificiality to destroy them."

Yet these Victorian owners and gardeners were highly successful on a basis of rule-of-thumb and traditional methods passed on by generations of glasshouse men. Since those times, no more than sixty or seventy years ago, all manner of revolutionary changes have taken place in these old rule-of-thumb practices. Indeed, had these quite revolutionary changes been geared to growing on the sumptuous scale of the nineteenth century, the present-day greenhouse must have been an amazingly colourful collection of superb beauty, unsurpassed in any age.

As it is the scientist and the research-worker have been probing into the odd corners of potting-sheds and under the dust of stagings, not to enable growers to grow and maintain a tropical splendour but to enable them to grow utility crops with ease and according to a set of rules both foolproof and of universal application.

Heating on scientific lines, a subject which had harassed the Victorians and on which they had spent hours of work and written scores of scientific articles, has entered a new era with the application of electricity to the problem. A simple formula which can give any owner, by an arithmetical process, the exact heat in British Thermal Units needed for any-sized houses and for any subjects to be grown in them, has taken the guess work out of that age-old problem.

The variety of efficient heating methods a gardener has at his command today is legion: hot air by electric fan and element; hot-water systems heated by electricity, coke, oil or gas; all-electric systems, electric strips, tubes, soil-heating wires, or cable—but (and here is the epoch-making difference) all these methods can be regulated automatically by thermostat to take both labour and chance out of heating, making it a precise instru-

ment in a gardener's hands. The veriest tyro can heat a greenhouse
to a given temperature and keep it at that temperature at a reasonable cost.

The new plastic films have also made tremendous changes in
the techniques of growing, giving the grower the advantage of
double-glazing—a Victorian innovation but a costly one then—
for very little indeed, with sheets of plastic used as underlining
for houses. The heat-retaining properties of such methods are
considerable and some authorities put them at between 5° F. and
15° F., while the reduction in heat loss is as high as 40 per cent.
With temperatures of 50°, 60° and 70° F., fuel savings on the
south coast in the month of December have been found to be as
high respectively as 8·8, 17·6 and 26·1 per cent.

For shelter-houses and as a far cheaper form of construction,
houses are being built completely of plastic film merely stapled
down to a light, uncomplicated wooden framework; points in
favour being extreme lightness (the house can be moved to cover
a cropping programme), since the film is one-sixtieth the weight
of glass, and extreme ease of erection. Snags still being ironed out
are the proneness of such a house to excessive humidity, and film
deterioration owing to ultra-violet light, involving fairly frequent renewals. Sheets of film are also being used in greenhouse
border work when, laid like a blanket on the soil and pierced at
intervals for plants to grow through, they asphyxiate all weeds at
birth. On benches the film is now being used as a "tray" in which
to put water, peat or gravel on which to stand or in which to
plunge the pots.

In propagation science has come to the gardener's aid with the
application of soil-heating wires and cables; there is also plastic
sheet draught-screening and seed-bed covering; and there are
mist-spraying techniques in which a finest overhead spray of
water, controlled by an automatic switching device which never
allows the bed to dry out, falls over a peat, sand or vermiculite
bed, keeping cuttings always moist, turgid and vigorous. This
latter practice, still being perfected, has already overcome many
of the difficulties and risks of the art. Many subjects previously

of the utmost difficulty to rear have proved amenable and easy to root under the system.

The question of light intensity and duration, although always a matter of intense speculation and theorising on the part of the nineteenth-century growers, has become precise and accurately calculable with the application of electric lighting to greenhouse cropping, and the stimulation of plant growth in the seedling stage has proved to be one of the quite spectacular applications of lighting to greenhouse cultivation. High-pressure mercury-vapour 400-watt lamps have proved to be of immense benefit. Batches of seedlings are placed under these lamps for 12-hour spells or, in the case of tomatoes, 18-hour spells.

With tomatoes experimental work has shown that an increase of up to 70 per cent in early crops and 10 per cent in total yield can be expected from using this source of light.

With flowers, gloxinias, "Charm" and "Cascade" chrysanthemums, antirrhinums, petunias, sweet sultans, China asters, tagetes and alyssum have proved to germinate well and grow on apace in the dark winter days under artificial daylight. Chrysanthemums are being made an all-the-year flower by a technique of artificial day and night. The chrysanthemum is, of course, a short-day flowering plant, but its flowering can be delayed by the use of additional lighting or advanced by shading, for a number of hours each day, so that growers have worked out a time–light schedule which gives blooms for each month of the year.

Bulb forcing has also gained from electricity applied to the greenhouse with under-staging or cellar-raising of bulbs, which are given 12 hours a day of 100-watt lighting per square yard.

Many research organisations have sprung up to work on glasshouse problems, not, it must be admitted, for the sake of the private owners, for a feature of both between-the-wars and post-war horticulture has been the increasing importance to the national economy, larder and home decoration, of the commercial glasshouse industry, and it is this large-scale acreage of profitable and utility glass they serve.

In this context there at once springs to mind the John Innes

Institute and its work on so many aspects of glasshouse cultivation. Their most outstanding and far-reaching discovery to
date, which has rendered obsolete the old abracadabra of potting
and made a laughing-stock of the ancient secret witchcraft
behind locked potting-shed doors, is that of a simple compost
for seed germination and growing and potting which works for
everyone, be he amateur or professional.

Some of the longest-held traditional techniques and rules of
the glasshouse man have been put to the test and found wanting
by these research stations working for the commercial glasshouse
industry. Naturally, in the long run all such research must benefit
the owner of the smallest greenhouse and already many of the
findings of the research teams can be followed, and are followed,
by the amateur in his own back-garden house.

The John Innes staff, for instance, have gone forward from
their basic research on compost to test and prove that firm or
loose potting, so much a feature, one way or the other, of the
old plantsman's catechism, is not as a rule of major importance.
Another shibboleth of the old professional foreman was that cold
water was harmful to seedlings—a can of aired water was always
in the house and woe betide any young apprentice who dared to
use anything else—has been abandoned after testing. No signi-
ficant results can be found for the aired-water treatment, and cold
water was not harmful, said the researchers.

The old gardener always wanted his potting-on or seed compost with the chill off too, to give his small charges a good send-
off in life, but that practice has little to recommend it, says the
scientist.

A wearisome practice, to which the old hand always adhered as
if his life depended on it, was progressive potting-on from thumb
size to the final six or nine, or even larger pot, depending on the
plant grown.

To have departed from this régime in the old bothy days
would have meant that the offending hand would have to find
another job; the practice was sacrosanct, part of the unwritten lore
of the greenhouse, yet the modern research man advocates seeds

sown in the pot in which plants are to flower without any plant-damaging, labour-wasting potting-on.

Hydroponics—the art of growing greenhouse crops without soil at all—is also a feature of modern practice. Instead of soil, the greenhouse owner fills his house with tanks of nutrient liquids with enough chemical feed to preserve and grow his plants to full maturity in a vigorous state. All manner of crops have been tried under this method—salad crops and pot plants, and all with success.

Experimental work on house shape as related to light-attracting quality has also been carried out and another bit of Victorian garden lore has been shown to be unscientific. In this context the best house for vigorous plant growth in winter, when greenhouses suffer most in this country from lack of light, will not be orientated, as tradition had it, from North to South but from East to West and will not be even-spanned with a 45 degrees slope, but unevenly spanned with the longest side facing south at an angle in the region of 50 degrees and the other broken into two angles of glass, the lower one being roofed at an angle at about 60 degrees and the top one about 30 degrees. The angles are still being investigated and research generally on this type of house is still being carried out not only at the glasshouse research stations run for the commercial growers, but by Government Horticultural stations.

Nowadays, apart from commercial growers, the newer methods of construction and cultivation are very much the concern of the research stations themselves, the botanical gardens of Kew, Wisley, Oxford, Cambridge and Edinburgh, and the larger local authorities such as the L.C.C., and the big cities and towns such as Birmingham, Manchester, Liverpool and Leeds.

At the two latter cities two great ranges, probably the only two establishments in this century to match in either size or magnificence the glories of a Victorian estate garden, have been built since the last war, and each has cost a total of some £60,000, which gives an excellent indication why such ranges are more and more the monopoly of such rate- or tax-assisted authorities and why the landed gentry can no longer even begin to think of such luxuries as part of the surroundings of their homes.

At the Leeds establishment many of the new techniques have been carried out. Low-pressure steam heating, with an automatic stoker, is maintained at a pre-determined temperature by motorised valves worked by a thermostat. The largest panes thought possible, 56 inches by 28, are used for glazing, in bars of aluminium alloy, the widest section of which is no more than an inch wide. The light intensity in fifteen houses 127 feet long by 27 feet wide is such that a light meter used outside and inside shows but a negligible difference in reading.

While we have not exactly covered in with glass whole gardens or cities, there are one-acre erections to be seen up and down the country harbouring salad crops, and we have erected greenhouses almost at the poles to give comfort and green sustenance to travellers in these harsh and frozen regions. A lean-to greenhouse was built on to the base huts of the Commonwealth Trans-Antarctic Expedition for Sir Edmund Hillary at Scott Base on the McMurdo Sound where mustard and cress, radishes and a few flowers were grown during the Antarctic summer from October to April. Heating was by means of an extension of the central-heating system of the base huts and also by the intense solar power. Every piece of glass, putty, wood and even the soil—300 pounds of it—had to be taken out from this country.

The first British post-war exploratory base at Marguerite Bay, Graham Land, had a lean-to also, which provided the men there with radishes, lettuces, mustard and cress as food and hyacinths which flowered and scented the mess-room delightfully for Christmas and were claimed to be the first ever to be grown in the South Polar regions.

In the Antarctic, too, at Kerguelen Island, the French have a separately sited greenhouse at the experimental base, heated with oil, and they have been experimenting with hydroponics for the growing of cereals including maize.

In the Arctic Circle at the Aklavick Anglican Mission Hospital a greenhouse constructed for the sake of the patients is quite

R

ambitious, growing sweet peas, tomatoes, peaches and grapes, while in Canadian Alaska at the Kotzebue Hospital cucumbers and other salad crops are grown for the inmates.

The modern cultivation of house plants and the consequent incentive to be more ambitious and grow the more exotic types indoors has brought back the Wardian case in its decorative and contemporary form exemplified by the present-day terrariums which can and do include under that name anything from an empty gold-fish bowl or a disused chemical carboy to an electrically heated, lighted and thermostatically controlled "Tropicater" in which all manner of tender plants can be kept in good health in the drawing-room.

The future of the glasshouse proper in this country, however, would appear to be more in the direction of obtaining a scientific automation-utility-crop than into one of extreme floriferousness.

At present, the scientist declares he can still see the bricks and mortar of the old orangeries when he looks at the present-day house in terms of the broader-than-need-be spars, roof bars, ridges and the sizes of window-panes and overlaps.

The functional house of the near future, says the scientist, will have walls no more than one and a half feet high, gutters at ground level, panes of glass at least 24 inches wide embedded in plastic material, metal glazing bars and structure of the thinnest material, with no internal props or trusses.

Professor R. H. Stoughton, D.Sc., formerly Professor of Horticulture at Reading University, in a recent "George Johnson" lecture gave his picture of a modern house:

"In the all-electric greenhouse of the future, the temperature, humidity, heat, light, air and the supply of water and food to growing plants will be controlled and carried out automatically by electrical appliances. I dare to predict," he continued "that the glasshouse of the future will be a sealed structure, perhaps of transparent plastic, into which is blown air automatically conditioned at the right temperature, humidity and carbon dioxide content. The plants will be grown in an artificial medium with automatic control of their water and nutrient supply and a strictly-controlled temperature régime. Natural daylight will be capable of automatic supplementation by artificial light of controllable spectral composition . . ."

Pest control can also be largely automatic with electrically timed and discharged pest-destroying sprays or vapours, while high-frequency treatment of diseases in bulbs and the interior of plants is receiving the attention of research workers who are also subjecting seed to the bombardment of radiation from radio-active isotopes to speed sporting and mutations to secure new varieties of many greenhouse plants.

One garden seer looking into the plastic wonderland of the future prophecies the plastic-house in place of the glasshouse—a shell of plastic like a great crystal balloon, he says, will be our house by the end of this century, probably moulded or pressed in one piece, certainly curvilinear, and for the bigger houses unit sheet construction of unbreakable clear plastic, the members of which will also be of clear, or if need be, semi-transparent plastic, replacing all bars, framework, doors or walls.

Less than a year ago at St. Louis in the United States was opened one of these covered gardens of the future, so futuristic that even the old traditional nomenclature has been dropped— Climatron is the new name.

This building covering over three-quarters of an acre is built in the shape of a huge dome, 175 feet in diameter and 70 feet high; is made entirely of aluminium tubing arranged in hexagonal pattern, lined by a layer of quarter-inch-thick plastic "glass" and cost $700,000.

From a central control panel is programmed a complex of fans, dampers, water-spray nozzles and two completely independent air-circulatory systems, which enables the gardener to maintain several entirely different climatic conditions under the one roof. While experimentation is going on all the time under these entirely new growing conditions, the Climatron may house an "Amazonian forest" of hot humid air, and a "Little Hawaii" representative of the cool days and warm nights of oceanic shores, while an Indian dry tropical flora grows in a warm-day and cool-night section, with a tropical mist forest with rain-drenched atmosphere in another section.

A tropical pool and bog section, containing the Victoria Regia water lily and other luxuriating water plants, has been designed with a plastic glass tunnel running through its centre so that the Climatron's visitors can walk under the pool and see the plants in growth.

The originator of the St. Louis Botanic Gardens was one Mr. Henry Shaw who was inspired to build his first greenhouse after seeing Paxton's Great Conservatory exactly one hundred years ago.

In spite of progress, the old days were grand days and days surely well worth recalling and recording before vivid memories of these fade and before those who had a part in them pass on with their memories, memories of a British glasshouse tradition which in its heyday was not surpassed anywhere in the world, for the era of the hot-house and conservatory with its long traditions and history matched a great era in the British tradition of gracious living when money was used to create beauty without thought to the cost.

Whether sufficient money, inclination, labour, time, or new plant discoveries will ever again make possible such a glorious outburst of floral novelty for art's sake to satisfy a whim, a highly developed aesthetic taste, or just as a gracious background to living, it is impossible to prophesy.

Already most of the great nurserymen who searched the world for new plants with which to please their noble patrons and built up tremendous businesses with world-wide connections have gone, and if their firms remain they no longer deal in any major way in exotic and stove-house plants. Why, even a catalogue listing, as they used to do in bewildering variety, the hundreds of hot-house plants, aquatics, orchids and almost any worthwhile greenhouse plant under the sun, is a thing of the past and an old one is hard to come by.

Old gardeners, if persuaded, can occasionally produce from the dimmer recesses of the potting-shed cupboard a soil-spattered and well-thumbed treasure of a plant list, well illustrated, ticked and noted with names that are but an echo in these herbaceous days. Keeping a firm hold of the catalogue old Adam will just allow the favoured one to glance at what the old-established nursery firms used to consider a good list, when masters insisted upon the latest novelty, regardless of cost, to grace conservatory, stove and greenhouse, almost before the ship which brought the new plants from their foreign homes was fast at anchor. The modern catalogues listing hot-house plants are few and are but a pale shadow of the former glories of their predecessors. There is not the money to start or maintain a sizeable and adequate hot-house collection in these days, the professional gardeners say. For the majority of gardeners and greenhouse men it is an old stove-house plant or two salvaged from a previous era, coddled and cosseted over the boiler or in the warm blanket of the propagating pit that has to make do as the poor, poverty-stricken reminder of the glorious exotic collections of the past. Given the money, however, there can be little doubt that garden-lovers would be attracted once again to the novelty and charm of the jungle on the doorstep.

There is a long range of ruined houses attached to a Victorian mansion in Yorkshire where after dinner the master delighted to walk. He would light his cigar in the library, open an oak door, climb a concealed stair and find himself in a vast conservatory. He would walk slowly on into the heathery, pass into the vinery, come to the orangery, the stove, the geranium house, the camellia house, and turn to saunter back again and join the ladies of the household in the conservatory at the far end, given over to shrubs. When he reached the library again he had finished his cigar and had not been in the open air for an inch of the way.

Today we are, if anything, more garden-proud than ever, our interests are as keen in both indoor and outdoor work, and many of us would be delighted to finish our after-dinner smoke strolling in the conservatory. Quite apart from the financial problem common to so many of making both ends meet, however, it would cost thousands of pounds today to start all over again, if it were possible to collect from their native jungle, high mountain, meadows or damp, sultry valleys the hundreds of unusual plants a Victorian nurseryman kept as a matter of course.

Indeed many of those plants could not be found when so many parts of the world are being "civilised," which so often involves the destruction of natural beauty and conditions, as jungle and woodlands are eaten up for timber for our paper- and pulp-making plants, while ecological differences are being brought about by irrigation, afforestation and all the tremendous upheaval which accompanies them.

So, leaving the twentieth century at its half-way mark, we can say that while research on greenhouse practice and technique has never been more active or widespread, and, if we follow the rules laid down, never was greenhouse cultivation less difficult, yet we must regret with deep disappointment the greenhouse wealth we have lost and wonder whether it will ever return.

Shall we have to make do with our terrariums, our 12 × 8s, our tomatoes and chrysanthemums in pale imitation of former glories, or will science as well as finding us new and vastly improved houses and skills also find us the means to build up and richly endow and stock new hot-houses and conservatories once again? I, for one, sincerely hope so.

BIBLIOGRAPHY

ABERCROMBIE, J.	*The Hot House Gardener* (1789).
	The Gardener's Pocket Journal (1786).
AITON, W.	*Hortus Kewensis* (1789).
AMHERST, A.	*History of Gardening* (1895).
ANDERSON, A. W.	*The Coming of the Flowers* (1950).
ANDERSON, J.	*Description of A Patent Hot House* (1804).
ANON	*On Heat and Cold of Hot Houses* (1756).
ANON	*European Parks and Gardens* (1853).
BACON, F.	*Of Gardens.*
BAINES, T.	*Greenhouse and Stove Plants* (1885).
BANKS, J.	"Some Hints respecting The Proper Mode of Inuring Tender Plants to our climate," *Hort. Trans.* Vol. I (1812).
	"On the Forcing Houses of the Romans," *Hort. Trans.* (1812).
BEAN, W. J.	*The Royal Botanical Gardens, Kew* (1908).
BLIGH, W.	*A Voyage to the South Seas* (1792).
BLOMFIELD, R.	*The Formal Garden in England* (1901).
BOYLE, F.	*About Orchids* (1893).
	Woodland Orchids (1901).
BRADLEY, R.	*New Improvements of Planting and Gardening* (1718).
	A Description of a Greenhouse contrived for The Good Keeping of Exotics (1718).
CHALMERS	*Biographical Dictionary.*
CLAUDIUS, J.	*Hints on The Formation of Garden and Pleasure Gardens* (1812).
COBBETT, W.	*The English Gardener* (1829).
COLUMELLA	*Works.*
COMMELIN, J.	"Management, Ordering and Use of Lemon and Orange Tree," *Trans.* (1683).

COWELL, J.   *The Curious and Profitable Gardener* (1730).
COX, E. H. M.   *Plant Hunting in China* (1945).
CRUICKSHANKS, A.   "Upon a New Mode of Applying Hot Water to Heating Stoves, etc." *Hort. Trans.*, 2 Series (1835).
CULLUM, D.   "Paper on Stoves," *Phil. Trans.* (1694).
CURTIS, W. (ed.)   *Botanic Magazine* (1801).
CUSHING, J.   *The Exotic Gardener* (1814).
DAUBENNY,   *Oxford Botanic Garden* (1850).
   C. G. B.
DEFOE, D.   *Journey Through England and Wales* (1726).
DENNIS, J.   *The Landscape Gardener* (1835).
DEOAGULIERS   *Fires Improved—A New Method of Building Chimneys* (1716).
DICKS, J.   *New Gardener's Dictionary* (1771).
DODOENS, R.   *A New Herbal or Historie of Plants* (1619).
DUTTON, R.   *The English Garden* (1937).
DYER, G.   *Pompeii, Its History, Buildings and Antiquities* (1876).
ELLIS, T.   *Gardener's Pocket Calendar* (1776).
ESTIENNE, C. and   *Maison Rustique* (1573).
   LIEBAULT, J.
EVELYN, J.   *The French Gardener* (1672, trans.).
                      Kalendarium Hortense (1676).
                      Diary (ed. Bray, 1818).
FAIRCHILD, T.   *The City Gardener* (1722).
FIELD and SIMPLE   *Memories of the Botanic Garden, Chelsea* (1878).
FIENNES, C.   *Through England on a Side-Saddle in the Reign of William and Mary* (1888).
FORTUNE, R.   *Three Years Wandering in the North Province of China* (1847).
GARTON, J.   *Practical Gardener and Gentleman's Directory* (1719).

GERARD, J.   *Herball* (1636).
GIBBS   *Technological Repository* (1828).
GIBSON   "Account of Several Gardens near London," *Archaeologier* (1709).
GOTHEIN, M. L.   *A History of Garden Art* (1928, trans.).
GOWEN, R. G.   "Observations Upon Glazing of Hot Houses and Conservatories," *Hort. Trans.*, Vol. III (1820).
GREEN, D.   *Gardeners to Queen Anne* (1956).
GRIEVE, P.   *History of Variegated Zonal Pelargoniums* (1868).
GROEN, J. VAN DER   *Den Nederlandtsen Hovenier* (1670).
HADFIELD, M.   *Pioneers in Gardening* (1955).
HANBURY, W.   *A complete Body of Planting and Gardening* (1770).
HAWKS, H.   *Pioneers of Plant Study* (1928).
HAYWARD, J.   "An account of A Steam Apparatus," *Hort. Trans.*, Vol. III (1820).
HAZLITT, W. C.   *Gleanings in Old Garden Literature* (1887).
HERBST, J.   *New Green World* (1954).
HESSE, H.   *Neue Garten-Lust* (1714).
HIBBERT, S.   *New and Rare Beautiful Leaved Plants* (1869).
                    The Amateur's Greenhouse and Conservatory (1888).
HILL, J.   *The Gardener's New Kalendar* (1758).
HILL, T.   *The Profitable Art of Gardening* (1579).
HIRD, DR.   *Tribute to the Memory of Dr. Fothergill* (1781).
HOOKER, J. D.   *Himalayan Journals* (1854).
                    Journal of Sir Joseph Banks (1896).
JAMES, J.   "The Theory and Practice of Gardening," *Trans.* (1712).
JEFFERY, R. A.   "The Horticultural Trade 1804–54," *R.H.S. Journal* (1954).

JOHNSON, G. W.   *A History of English Gardening* (1829).

KENNEDY, J.   *Treatise Upon Plants and Gardening and Management of the Hot House* (1776).

KENT, W.   "Some Further Account of the Management of a Stove for Tropical Plants," *Hort. Trans.*, Vol. III (1820).
"Account of Some Improvements in the Construction of a Stove for Plants," *Hort. Trans.*, Vol. II (1818).

KNIGHT, A.   "On the Ventilation of Forcing Houses," *Hort. Trans.*, Vol. II (1818).

LA COURT, PIETER DE   *Byzondere Aenmerkingen over het genleggen van pragtige en gemeene landhuizen* (1737).

LA QUINTINYE, J. E.   *The Compleat Gardener* (1693).

LANGFORD, T.   *On Fruit Trees* (1696).

LANGLEY, B.   *New Principles of Gardening* (1728).

LAURENCE, J.   *The Clergyman's Recreation, Showing the Pleasures and Profit of the Art of Gardening* (1717).

LAWRENCE, W. J. C.   *Science and the Glasshouse* (1950).

LAWRENCE and NEWELL   *Seed and Potting Composts* (1944).

LAWSON, W.   *New Orchard and Garden* (1618).

LETTSOM, J. C.   *Hortus Uptoniensis* (1774).

LEVI, D.   *Wages and Earnings of the Working Classes* (1867).

LIGHTOLER, T.   *The Gentleman and Farmers Architect with plans of Greenhouses, Pineries, etc.* (1762).

LIVINGSTONE, J.   "Observations on the Difficulties which have Existed in the Transference of Plants from China," *Hort. Trans.*, Vol. III (1820).

LODDIGES, C.   *Botanical Cabinet* (1824–27).
LONDON, J. and   *The Retir'd Gardener* (1706).
  WISE, H.
LOUDON, J. C.   *Treatise on Several Improvements Recently
                        Made in Hot Houses* (1805).
                      The Greenhouse Companion (1824).
                      An Encyclopaedia of Gardening (1825).
                      *The Suburban Gardener and Villa Com-
                        panion* (1838).
                      Self Instruction to Young Gardeners (1845).
                      The Villa Gardener (1850).
LOWE, E. J. and   *Beautiful Leaved Plants* (1866).
  H. W.
MCDONALD, A.   *A Complete Dictionary of Practical Gar-
                        dening* (1807).
M'GRATH, R. and   *Glass in Architecture and Decoration* (1937).
  FROST, A. C.
M'INTOSH, C.   *The Greenhouse, Hot House and Stove* (1838).
                      *Flora and Pomona, or the British Fruit and
                        Flower Garden* (1838).
                      The Book of the Garden (1853).
MACKENZIE, G. S.   "On the Form Which the Glass of a
                        Forcing House Ought to Have," *Hort.
                        Trans.*, Vol. II (1818).
MADDOCK, J.   *Florist's Directory* (1822).
MALCOLM, W.   *A Catalogue of Hot House and Greenhouse
                        Plants* (1771).
MANGLES, J.   *The Floral Calendar* (1839).
MARKHAM, V.   *Paxton and the Bachelor Duke* (1935).
MARTIAL   *Works.*
MAUND, B.   *Botanic Garden* (1825–32).
MAWE, T.   *Every Man His Own Gardener* (1717).
MILLER, P.   *The Gardener's Dictionary* (1731).
                      Garden Kalender (1745).
MILLICAN, H.   *Travels and Adventures of an Orchid
                        Hunter* (1891).

MORIARTY, H. M.   *Fifty Plates o, Greenhouse Plants* (1807).

MUIJZENBERG,   *Summary of Architectural Development of*
  E. W. B.   *Hot Houses* (1946).

NICHOLSON, G.   *The Illustrated Dictionary of Gardening*
                       (1844–88).

NICOL, W.   *The Scotch Forcing and Kitchen Gardener*
                       (1798).

PARKINSON, J.   *Paradisi in sole* (1629).

PAXTON, J. and   *Paxton's Flower Garden* (1850–53).
  LINDLEY, J.   *Magazine of Botany* (1834–49).

PEPYS, S.   *Diary* (1666).

PERFECT, T.   *Practice of Gardening* (1759).

PLAT, H.   *The Garden of Eden* (1675).

PLINY   *Works.*

PULTENEY, R.   *Historical and Biographical Sketches of*
                       the Progress of Botany in England
                       (1790).

REA, J.   *Flora, Ceres and Pomona* (1676).

REES, A.   *Encyclopaedia of Arts, Science and Litera-*
                       ture (1819).

REPTON, H.   *Theory and Practice of Landscape Garden-*
                       ing (1805).

RHIND, W.   *History of the Vegetable Kingdom* (1857).

RICHARDSON, R.   *Correspondence of Dr. Richard Richardson*
                       of East Bierley (1835).

ROBERTSON, W.   *A Collection of Various Forms of Stoves*
                       Used for Forcing Pines, Fruit Trees and
                       Preserving Tender Exotics (1798).

ROBINSON, W.   *The English Flower Garden* (1889).

ROHDE, E. S.   *The Old English Herbals* (1922).
                   The Old English Gardening Books (1924).

SABINE, J.   "Observation on the Glazing of Glass
                       Houses," Hort. Trans. Vol. IV (1822).

SENECA   *Works.*

SERRES, O. DE   *Le Théâtre d'Agriculture et Mesnage des*
                       Champs (1663).

SHARROCK, R.   *The History of the Propagation and Improvement of Vegetables by the Concurrence of Art and Nature* (1660).

SHAW, C. W.   *The London Market Gardens* (1880).

SHAW, J.   *Plans, Elevations and Sections, with Observations and Explanations, of Forcing Houses in Gardens* (1794).

SHELDRAKE, T.   *Gardener's Best Companion in the Greenhouse* (1756).

SIEVEKING, A. F.   *The Praise of Gardens* (1899).

SIMMONDS, A.   *A Horticultural Who Was Who* (1948).

SMITH, E.   *The Life of Sir Joseph Banks* (1912).

SMITH, J.   *Historical Record of the Royal Botanic Garden, Kew* (1880).

SMITH, W.   *A Dictionary of Greek and Roman Antiquities.*

SPEECHLY, W.   *A Treatise on the Culture of Pineapples* (1779).

STEELE, R.   *An Essay Upon Gardening* (1800).

STEVENSON, H.   *Gentleman Gardener* (1766).

STEWART, H.   *The Royal Family of Plants* (1885).

STOTHERT, H.   "Description of Various Modes of Heating by Steam for Horticultural Purposes," *Hort. Trans.*, Vol. I, 2nd Series (1853).

STOUGHTON, R. H.   "Electricity in Efficient Horticulture," Reading University Paper.

SWEET, R.   *The Hot House and Greenhouse Manual* (1826).

SWITZER, S.   *The Nobleman, Gentleman and Gardeners' Recreation* (1715).
                      Ichnographia Rustica (1718).
                      The Practical Fruit Gardener (1724).

TAYLOR, G.   *Some 19th Century Gardeners* (1951).
                      Victorian Flower Gardens (1952).

TEMPLE, W.   *Upon the Gardens of Epicurus* (1908).

THEOPHRASTUS   *Works.*

THOMPSON, F.   *A History of Chatsworth* (1949).

THOMPSON, J. W.   *A Practical Treatise on the Construction of Stoves and Other Horticultural Buildings* (1838).

TOD, G.   *Plans, Elevations, and Sections of Hot Houses, Greenhouses, and Aquariums and Conservatories, etc.* (1807).

TROWELL, S.   *A New Treatise of Husbandry, Gardening and Other Matters Relating to Rural Affairs* (1739).

VAN OOSTEN   *The Dutch Gardener* (1703).

VARIOUS   *Florist's Manual* (1822).

VEITCH, J.   *Manual of Orchidaceous Plants* (1887). *Hortus Veitchi* (1906).

WALLICH, N.   "Upon the Preparation and Management of Plants During a Voyage from India," *Hort. Trans.*, Vol. I, 2nd Series (1835).

WARD, F. KINGDON   *The Romance of Plant Hunting* (1924).

WARD, N. B.   *The Growth of Plants in Closely Glazed Cases* (1842).

WATSON, W.   "An Account of the Bishop of London's Garden at Fulham," Royal Society Papers.

WESTON, R.   *Tracts on Practical Agriculture and Gardening* (1769).

WETHERED, H. A.   *A History of Gardening* (1933).

WHATELY, T.   *Observations on Modern Gardening* (1770).

WILKINSON, T.   "Observations on the Form of Hot Houses," *Hort. Trans.*, Vol. I (1818).

WILLIAMS, B. J.   *Choice Stove and Greenhouse Plants* (1869). *Choice Stove and Ornamental Leaved Plants* (1870).

WORLIDGE, J.   *Compleat System of Husbandry* (1716).

INDEX

271

S

Plant Propagation in Pictures

THIS book contains the complete story of plant propagation by accepted modern methods, written in a delightful "at-your-elbow" style. Each chapter consists of a short introduction giving the basic principles, followed by nearly 400 step-by-step pictures with fully explanatory captions.

Plant Pruning in Pictures

No serious gardener can afford to be without this book, which covers every aspect of pruning, including advice on when *not* to prune and why. The clear and straightforward text is illustrated with nearly 400 drawings and photographs showing step-by-step procedures for pruning both woody and herbaceous plants, annual and perennial flowers, fruits, herbs, evergreens and deciduous trees and shrubs, and house plants.

Each volume Large Demy 8vo   *32s. 6d. net*